SCIENCE, CHURCHILL AND ME
The Autobiography of Hermann Bondi

Related Pergamon Titles of Interest

GOTSMAN
Frontiers of Physics (Proceedings of the Landau Memorial Conference,
Tel Aviv, Israel, 6–10 June 1988)

KHALATNIKOV
Landau: The Physicist and the Man

LUTHER
Advances in Theoretical Physics (Proceedings of the Landau Birthday
Symposium, Copenhagen, 13–17 June 1988)

PITAEVSKI
Collected Papers of E. M. Lifshitz

TER HAAR
Collected Papers of P. L. Kapitza Volume 4

PROFESSOR BONDI

SCIENCE, CHURCHILL AND ME

The Autobiography of
Hermann Bondi,
Master of Churchill College Cambridge

PERGAMON PRESS

Member of Maxwell Macmillan Pergamon Publishing Corporation

OXFORD · NEW YORK · BEIJING · FRANKFURT
SÃO PAULO · SYDNEY · TOKYO · TORONTO

U.K.	Pergamon Press plc, Headington Hill Hall, Oxford, OX3 0BW, England
U.S.A.	Pergamon Press Inc., Maxwell House, Fairview Park, Elmsford, New York 10523, U.S.A.
PEOPLE'S REPUBLIC OF CHINA	Pergamon Press, Room 4037, Qianmen Hotel, Beijing, People's Republic of China
FEDERAL REPUBLIC OF GERMANY	Pergamon Press GmbH, Hammerweg 6, D-6242 Kronberg, Federal Republic of Germany
BRAZIL	Pergamon Editora Ltda, Rua Eça de Queiros, 346, CEP 04011, Paraiso, São Paulo, Brazil
AUSTRALIA	Pergamon Press (Australia) Pty Ltd, PO Box 544, Potts Point, NSW 2011, Australia
JAPAN	Pergamon Press, 5th Floor, Matsuoka Central Building, 1-7-1 Nishishinjuku, Shinjuku-ku, Tokyo 160, Japan
CANADA	Pergamon Press Canada Ltd, Suite No 271, 253 College Street, Toronto, Ontario, Canada M5T 1R5

First edition 1990

Library of Congress Cataloging in Publication Data

Bondi, Hermann, Sir.
Science, Churchill, and me: the autobiography of
Hermann Bondi, master of Churchill College,
Cambridge.
p. cm.
1. Bondi, Hermann, Sir. 2. Scientists—Great
Britain—Biography. 3. Mathematicians—Great
Britain—Biography. I. Churchill College. II. Title.
Q143.B56A3 1990 509.2—dc20 90–6997

British Library Cataloguing in Publication Data

Bondi, Sir, Hermann *1919–*
Science, Churchill and me: the autobiography of
Hermann Bondi, Master of Churchill College
Cambridge.
1. Scientists. Biographies
I. Title
509.24

ISBN 0-08-037235-X

Contents

Foreword

In the 1930s a number of Austrian and German men and women left their country in the face of personal danger and racial discrimination. Some very obvious names come to mind; Lord Weindenfeld, Sir Claus Moser and many others, In the years that followed their intellect, skills and talents enhances and enriched the life of this country. The name of Hermann Bondi stands high on that roll of honour.

Most of those who read this fascinating account of his life will never have had to face those problems. It is difficult for us to imagine the circumstances in which a boy had to leave his country and finds himself in an alien land continuing his education in a foreign tongue. Not many of us would have managed to get to Trinity College, Cambridge and fewer still become a Fellow a few years later or indeed made the resounding success of our lives as Hermann Bondi has done – to the great benefit of his fellow citizens. I first met Hermann Bondi over thirty years ago whilst I was living in Canberra. He had then as now a formidable reputation as a mathematician and astronomer. To someone with no scientific education he threatened to be a daunting guest: far from it. He proved to be one of the most engaging, humorous and brilliant people that I have ever had the privilege of meeting. He has the gift of explaining the most complicated proposition in terms which a layman can understand, without condescending to those much less intelligent than himself.

The next time that our paths crossed I was looking for a successor as Chief Scientist at the Ministry of Defence and Sir Solly, now Lord Zuckerman, suggested his name to me to have exactly the qualities required in a job which demands a brilliant scientist but at the same time a man who can explain to politicians and servicemen the most difficult and increasingly baffling problem. Not, I think, without some hesitation Hermann Bondi accepted. He was an instant and outstanding success. Those who worked with him understood and enjoyed him and above all respected his judgement.

And so it was in all his many appointments before and since. Churchill College has been fortunate to have him as its Master. I am sure that the Great man after whom it was named would have looked with approval on his tenure of office and been gratified that so distinguished a man should be its Master.

The Rt. Hon. Lord Carrington, K.G.

Preface

THE ONLY excuse for writing an autobiography is that one thinks one wants to convey something of interest to others. In my case it is amazement that I should have had so varied and fascinating a career, that I enjoyed meeting so many marvellous people in very different walks of life, and that so absorbing and enjoyable a collection of jobs should have come my way. Sometimes it seems to me that I have been walking through life with a wide open mouth, and roast ducks have come flying in with monotonous regularity.

I have never had a particularly high opinion of my intelligence and have told my children, only half in jest, that it was much more important to be lucky than intelligent. They are quite ready to believe that I am lucky!

The reader will readily appreciate the cardinal importance to my life and indeed existence of the Second World War, Winston Churchill's leadership and its victorious outcome. Therefore, my pleasure in writing this in Churchill College is well founded.

CHAPTER 1
Vienna

ONE OWES a very great deal to one's home and school background, to one's friends, to one's parents, and to other relatives. I therefore need to give a reasonably full account of where I came from. My paternal grandfather came from Mainz in Germany. He was in business but not very successfully, and found it a good idea to move in 1884 to Vienna, to try to make a living again in business there. He was Orthodox as most Jews were at the time and had an enormous family of sixteen children, all children of my paternal grandmother whom I never knew. Two of those sixteen I believe, died very young, and there were two pairs of twins amongst them, and my father and his twin sister, born in June 1878, were the thirteenth and fourteenth in the family. I did indeed have an enormous number of first cousins on my father's side, and of all that generation I was the youngest bar one. I am pretty sure I never met more than a fraction of my cousins, since many lived in other places and, of course, were much older and died quite a while ago. Virtually by tradition, the first son became a Rabbi, the second one (who from all accounts was an extremely vigorous and energetic man) effectively took over the business, made it do rather well, but died fairly young early in this century. My grandfather, I should say, who was born in 1831, lived to his mid-nineties and although there was over 88 years between us I can still remember him quite well. The other family with which I am so closely connected bears the name of Hirsch. They were well-to-do in the metal business and lived in Halberstadt, in what is now East Germany. There are many marital connections between these families. My father's mother, my mother's father and my mother's mother were all of the Hirsch family, as was a favourite aunt of mine, married to an older brother of my father's. The first link seems to have arisen from my paternal grandfather in his energetic young days, having struck a very successful arms deal with Garibaldi, travelling as was necessary in those days by coach over the St. Gotthard pass. This I was told led to the Hirsch family accepting him as a son-in-law.

My own father had the temperament and outlook of a scientist but, of course, there was not much in the way of career possibilities for scientists in those days, even less so for a Jew, and he decided to become a doctor. But all his life he had a hankering for the scientific side of medicine in

1

which he had done significant research work as a young man at Heidelberg; in his older days he tackled the difficult problem of the origin of the different noises made by healthy and diseased hearts. His scientific inclinations had a very strong influence on me. On every occasion where there was something of interest to see he stressed how remarkably interesting were the questions that arose about the origin of this or the nature of that. This attitude of curiosity has certainly rubbed off on me. A somewhat older brother of his, Joseph, also became a medical doctor and this was the branch of the family with whom we had the closest relationship. Joseph's wife Rosa was another Hirsch, an aunt of my mother's, who had introduced my mother and father to each other when my mother was on a visit to her aunt in Vienna. Thus these two families have always been most closely connected. Though my mother's family were very well-to-do, there was a great deal of tragedy in their lives. My mother was the second of two children, her brother some years older than she. Having been terribly ill with meningitis as a child, he never fully recovered his mental equilibrium. Her own mother died in childbirth, perhaps not too rare an occurrence in those days (1892). Her father then married his first wife's sister and out of that union there sprang my Uncle Seppi, a great favourite of mine, six years younger than my mother. My grandfather's second wife too died very young, of tuberculosis, and he himself did not survive her very long. Indeed, of all my four grandparents, only my paternal grandfather lived into my own lifetime. My mother, very intelligent but with only limited education as was not uncommon for girls in a small German city in those days, again always had a very questioning outlook. She invariably took the greatest interest in medical, educational and political questions, and had a real intellectual partnership with my father, in spite of the deficiencies of her education. She knew to perfection how to control her naturally protective instincts *vis-à-vis* her children, and encouraged our independence from early on. She was, however, inclined to pessimism. She used to say on our walking holidays in often rainy areas that the weather was the only issue on which she was an optimist. However, her pessimism was in no way aggressive until her last years. Indeed she was a splendid companion for us. My parents married in Vienna in May 1914, just before the First World War, which led to a long separation between my parents and a great worry for my mother, since my father as a junior doctor was in very exposed positions in the early stages of the war. Later he was ordered to run a hospital for prisoners of war (mainly Russian) amongst whom typhus was raging. This was in a remote part of Austria and with great courage he brought his wife and little daughter, my sister, to live there, being quite convinced that if their water supply came from above the camp and the hospital, there was no danger to them. My sister, born in early spring 1915, was understandably not followed by another child until I was born on 1 November 1919, completing the family.

There was an interesting contrast in the attitude to religion between my parents. Neither of them were believers, but where my father liked to follow the forms because of their value as tradition and as social cement (many decades later I was interested to learn from Sir John Colville that Winston Churchill had felt much the same), my mother was from very early on a strong rebel against the family orthodoxy and made it clear how much she disliked the narrowness, the self-satisfaction, the blindness of religion. What horrified her was, for example, the rule in Jewish Orthodox practice that if a child of yours leaves the faith, you should mourn for it as though it had died. Of course this dislike of orthodoxy had to be somewhat hidden because so many of the rest of the family were firm believers. Of the three brothers and one sister of my father who lived in Vienna, one was exceedingly orthodox with a family of four children, one fairly so with three. The other medical man, Joseph, shared my father's views and attitudes, whereas the sister, my aunt Lea, had very much the attitude of my mother. To keep relations with the rest of the family going one had of course to engage in a certain amount of pretence. Thus we were mildly kosher in our eating at home, but in no way when we were away on holiday. This enabled us to invite other members of the family to come to us for meals, though they might be occasionally a little mistrustful of the strictness of our cooking arrangements. My aunt Lea was a considerable personality, passionately interested in art. An art dealer of great distinction she was instrumental in popularizing Kokoschka and introduced Picasso's work to Vienna. She too had a rather peculiar life. I mentioned earlier on the second son of the family who ran the business, and then died rather young. He left four little daughters with their young mother (also a Hirsch) who was widowed so early in life. In the normal manner of the day the youngest available sister, Lea, was asked to help with the education of the four girls, supporting their mother in this. Naturally those links between the four and her remained intensely strong. Although this took some of her time and energy, yet she carved out for herself a position and standing that was all her own, and her own wide circle of friends, very much of artistic inclination and ability. Amongst them was a gifted sculptor, quite a bit older than herself, with whom she started a long-lasting relationship. He was of Jewish origin but Catholic and married, so for a long time there was no question of marriage. Eventually his wife, who was, I believe, a little older than he, died. Then at last she could marry the love of her life, a few years before emigration. Her husband sadly did not live for very long, succumbing to a kidney disease during the war years. She herself lived well into old age and died only in 1971. All through these years meeting her was a stimulating, interesting and instructive experience every time.

In my days as a little boy I was a worry to my mother because I had no great interest in food, and was, I gather, painfully thin for quite a few years. I also had repeated trouble with my ears and though this led to no lasting damage it kept me out of school for quite a while when I was about 6. Though private schooling was not nearly as common in Austria as in England, it was not at all unusual amongst our circles. My mother however was firmly opposed to it and both my sister and I went to ordinary state or at least state-supported schools all the way through. I don't remember much about my years in primary school but I do recall that I was already very firm about who my friends were going to be and that this was not going to be dictated to me by parents, where I was soon able to show a distinct dislike for the son of a friend of my mother. Certainly my school life was interrupted with very frequent colds, flu, sore throats and the like. Taking tonsils out was very common in those days but my father, as a physician generally not keen on surgery, did not wish this to be done in my case until I was a little older. I think it is fair to say, whether through the operation or through the passage of time, after this was done (about the time I was 12) I became much tougher and more energetic. This may have had a connection with my relations with the other boys in my school. Whereas in my early years even at secondary school, I only had a very small circle of friends, this changed very sharply when I was 13 or 14 and suddenly I was on the best of terms with everybody and very much liked throughout the class. I myself very much enjoyed company.

My sister was quite an influence on me in many different ways. In spite of the difference in age we both greatly enjoyed the Greek myths, and a good deal else in literature. It is very difficult to point in memory to definite events, but I must have been good at mathematics already when I was 8 or 9. My sister, for whom this was not the strongest side in spite of her generally scientific inclination, asked me to help her with her school work in mathematics when she was perhaps in the second or third year of her secondary school. I think I could be quite effective even then. I thoroughly enjoyed looking at her textbooks but the real turning point came when on a visit to uncle Joseph, I noticed a book (perhaps when I was 12), which was a simple introduction to calculus. In spite of being generally very encouraging, neither my uncle nor my parents thought that I would be able to follow this book, but in fact I swallowed it hook, line and sinker, and found nothing in it of any great difficulty. It is true of course that, as a youngster would, I became much more expert at manipulation than at the difficult logic of analysis. But I certainly did not find anything difficult in the book. At my own secondary school in mathematics I was far ahead of what we did in class. I remember class examinations when there were usually two sets of questions to be given to alternate boys to avoid cheating. In an exam lasting perhaps two hours, within twenty minutes I had generally answered every question on both papers. Though there was the normal opposition to swots in class, the patent effortlessness of my mathematical successes meant that this caused no hostility at all. I also became rather competent at the spoken word and at hiding from teachers any ignorance that arose from my unwillingness to spend too much time on work that did not interest me all that much. This also applied to examinations, many of which were oral. When I was about 15 I acquired an undergraduate text on theoretical physics by Joos (which I believe is still quite popular and which certainly helped me greatly). Some parts of it were evidently beyond me but much of it I could absorb relatively easily. My interests on that side continued to evolve without other things being excluded entirely. At this moment I should also refer to a tragedy that occurred: about 1927 my uncle Joseph and aunt Rosa lost their only daughter, Helen, a talented pianist at the beginning of a successful professional career, to an attack of pneumonia which of course was a terrible disease in those days. Their son, Arthur, took a particular interest in me and in particular my social education in matters like wine and interests like politics. The two families became even closer after this sad event.

But having mentioned the word politics, I must enter upon the difficulties and unpleasantness of the time which has had a strong influence on my outlook all my life. Austria in those days was a very divided and very poor country, with an overall population of barely six million. Nearly two million lived in Vienna. The other cities were by comparison quite minor. The rural population dominated in all the parts of the country (which was federal) except Vienna, though in the north of Styria there was another significant industrial area. Industrial areas inevitably voted for the Social Democratic Party whereas in the rest of the country there was an easy majority for the Christian Social Party, very conservative and much dominated by the Catholic clergy. Inevitably this made the Social Democrats very anti-clerical, which fitted in with their general attitude. This party, while sharply separated from the Communists, was very much on the Left of what elsewhere might be called Social Democratic. In the upshot through the twenties and early thirties, the Christian Social Party had a small but sufficient majority to govern. The Social Democrats ruled Vienna, had an influence in Styria but were elsewhere fairly insignificant. There was a small Liberal Party of relatively little consequence. The political tensions, with two parties not properly on speaking terms mirrored the social tensions. The rural areas fed themselves, if not always well, the industrial areas were inevitably very poor and poorly housed. A major source of conflict was the imposition by the Social Democratic Government of Vienna of substantial taxes on properties commanding higher rents, from which they gained enough money for a large public housing programme. Though the forms of parliamentary democracy were largely observed, both parties built up para-military forces of their own, the Heimwehr for the Christian Social Government Party, the Schutzbund for the Social Democrats. In 1927, when I was still a small boy, this led to pretty bloody riots in Vienna but then things calmed down a little until the World Depression of the early thirties aggravated tensions everywhere. Even before Hitler came to power on 30 January 1933 there was a pan-German movement. Indeed these sentiments had been very widespread and when the Social Democrats had come to power immediately after the ceasefire of 1918 and the breakup of the Austro–Hungarian monarchy, the obvious idea was that the German speaking parts of the old Austria–Hungary should join the new republican Germany, and nobody in Austria disagreed with this. The Allies thought otherwise and, given the very turbulent time in Germany in the early twenties, the appetite for linking up diminished. Amongst the Catholic Christian Social Party this occurred because Germany was not a Catholic-dominated country, among the Social Democrats, because of the strength of the anti-Republican anti-constitutional forces in Germany. With Hitler's rise to power, and indeed just before, there began to be a growth in the pan-German ideas coupled of course with an increase in the Nazi Party, which previously had been rather small. The Christian Social Government found the parliamentary forms inconvenient and their lawyers found a law passed in 1917 and never repealed, that during the extreme economic difficulties caused by the world war, the Government could govern by decree. To stretch this to the conditions of 1933 was perhaps a little far-fetched but not necessarily illegal. A row then occurred in Parliament of such intensity that the Speaker resigned during the meeting and, not sure what they should do, his two deputies resigned as well, whereupon the meeting broke up and the Government, again using a legal subterfuge, claimed that since the sitting had never ended, not only could the main Parliament not meet because it had no chairman, but since it was unheard of for committees to sit while the whole of the House was sitting, it stopped committee meetings as well. And so, in the early spring of 1933, Austria had become a dictatorship. At first this was not oppressive but soon in the eight Federal provinces outside Vienna, the para-military force of the Government made sure that the local Parliament also failed to meet. The question was then in everybody's thoughts, what would happen in Vienna? Eventually, on 12 February 1934 in a manner that never became wholly clear, the Social Democratic Party called a General Strike and brought out their para-military force. It turned out that their housing projects had been very strategically placed and could control all the main railway lines into Vienna. Fighting was heavy and I can still recall the sound of artillery in the city which was used ruthlessly to destroy those blocks. After several days the fighting ceased and now the country was really in the grip of a brutal dictatorship, clearly only supported by a modest minority because by this time the rising star of Hitler had attracted many people to the Nazi Party, making the Christian Social Party almost certainly a distinct minority. Sure enough, in the summer of 1934, the Nazis started a second civil war which the Government, however, was able to contain and suppress. Externally Austria's independence was wanted by all the powers other than Germany, that is Britain, France and Italy, all still closely allied. Italy, ruled by Mussolini was of course specially keen to keep Germany away from the Brenner Pass. Through particularly Italian support the Government was able to withstand German pressures and suppress the rising of July 1934, and so Mussolini was seen as the great protector. I, myself, like almost any boy, horrified at the social tensions, horrified at the brutality of the suppression of the February 1934 rising, was strongly to the Left in my sympathies, as indeed, was my mother. We had always taken the Social Democratic Party paper as the newspaper we read at home, but many other people in our professional, and largely Jewish circles, felt not only, and rightly, that the Nazis were the great risk, but also that Austrian independence relied entirely on Mussolini, and were therefore prepared tacitly rather than overtly to support the Government and favour Mussolini. The situation looked very stable until Mussolini's Abyssinian ambitions, and the great hostility this aroused in the United Kingdom and France, split Italy off from her old alliance with the Western Powers. This development led to the formation of the axis and greatly undermined the external situation on which the government of Austria relied.

It was against this background that I threw myself into my mathematical physics and dreamt of going to live elsewhere. A country that had presented one with two civil wars within six months did not seem to me the happiest of places to be. However, I cannot claim that I had the great foresight to see that the stability of Austrian independence was not going to last. Oppressive the atmosphere certainly was and nothing was more comforting than a holiday in Switzerland, with its wonderful open democratic traditions. Czechoslovakia too was an outpost of democracy and once or twice we went there for holidays. Though one went occasionally to Italy, particularly to the little peninsula of Istria which is now Yugoslav, one was well aware of the nature of Mussolini's rule. I remember one had to be careful not to wrap one's shoes for the journey in pieces of the Social Democratic newspaper which would be strongly disliked by the Italian Customs if they looked through one's baggage. Of course, the south Tyrol, with its tensions between the German-speaking population and the Italian Government, was also an area one went to at times, but this was not nearly as enjoyable and recuperative as Switzerland with its splendid democratic climate. Holidays altogether played a considerable part in my life and thinking. I had a deep interest in walking and climbing mountains, which I did on every available opportunity, and also in skiing, which on occasions could be done fairly close to home but better further west in Austria where every year I went with the whole school class for a week. I became very knowledgeable about the topography of the Alps and particularly about the magnificent engineering of trans-Alpine railways wherever they were to be found. Altogether I found railways very fascinating, not so much like so many boys who like to sight railway engines, but the routing of railways, their timetables, the complexities of trains with through-carriages to many different parts of the Continent, and so on. To this day I like to see an international train with carriages with many different destinations, and marvel at the intricacy of a system which can make this work, albeit at very modest speed in places. In one's last year at school, as part of the final exam, one could choose to write a modest thesis, for which I chose the Alpine railways of Switzerland, exploring them from travel guides and timetables where I did not know them from my own experience. The finest pieces of engineering like the Albula line from Chur to the Engadin, or the Gotthard line, to this day give me the greatest pleasure and excitement as they did in my boyhood. Nearer to Vienna, that magnificent first railway crossing of the Alps, the railway running southwest from Vienna over the Semmering, with numerous tunnels, viaducts and curves, was a great source of inspiration to me. So in spite of the disagreeable political situation I had plenty to please me and enjoy myself with.

Perhaps one incident of that period might be worth recording. I belonged to the local Boy Scouts and we had occasional camps, including in July 1936 a camp on a tiny Dalmatian island in Yugoslavia, not far from the bigger island of Rab, all very lonely, very beautiful and rather good fun (I must confess though that then, already, I began to develop a liking for a lush landscape like Alpine forests and meadows, rather than the bare hills of the Dalmatian coast). I had to get home a little before the camp ended to join my parents on a trip. So, early one morning, the other boys rowed me across the glassy Adriatic to the city of Rab, where perhaps three times a week a steamer called early in the morning, that would at lunchtime reach Yugoslavia's northernmost port, Susak, then the sole Yugoslav part of what is now Rijeka. Susak was separated by a barbed wire barrier from Italy. This barrier which I had seen from both sides, made a deep impression on me. On arrival at Susak, before boarding the train for the 16-hour journey to Vienna, of course I wanted to get a newspaper, not having had any news for a long time. Naturally I asked at the international newspaper stand at the station, first for a paper in German. "Sorry sir, today they've all sold out." Then in English, "Sorry sir, today they've all sold out." This of course only whetted my appetite because clearly something had happened, and I went through the languages hoping to learn something, but in fact the only newspaper they had got left was a Serb paper in Cyrillic script. I told myself that on this boring train journey my mathematical knowledge of the Greek alphabet would allow me to decipher this and so it did. The Spanish Civil War had broken out.

Though I was never good at ball games and had been very thin and feeble as a little boy, in my teens I was pretty tough and determined. In school skiing parties it was often left to me to be the last one down the slope, so I could make sure nobody had been left behind. But in practical work I lacked any desire for precision. It may come as a surprise to people that somebody with a mathematical inclination is not favourably disposed towards precision, but I have always liked the idea of experiments and explanation that could be qualitative rather than dependent on the exact figures, that could be sketched out rather than drawn with perfection, and yet convey the information. This attitude did not make me a great favourite with relevant teachers but allowed me to grasp the essence of subjects relatively quickly. I also developed a good facility with my speech so that, particularly in oral examinations, I could easily convey an impression of much greater knowledge than I actually possessed. Then already I was beginning to be good at thinking on my feet, perhaps a little better than when sitting down.

CHAPTER 2
The move to England

IN THE summer of 1935, two years before the completion of my school career, we (my mother, sister and I) went to the Upper Engadine for our holiday. It is a most beautiful and intriguing area, a broad, smiling, gently sloping valley containing a chain of lakes surrounded by high mountains, with lovely forests and an excellent system of footpaths. Though marginally affected by the big boxes of opulent hotels in St. Moritz and to some extent Pontresina, it is yet a most attractive area. However, in that summer there was a cloud of worry, for there was a distinctly more than average outbreak of polio in Eastern Switzerland. It is hard to imagine, decades after the Salk–Sabine revolution effectively eliminated the disease, what a frightening scourge this crippling killing disease was in those days. Our mother therefore decided that we should leave the Engadine and continue our holiday in another, far less affected, part of the Alps, the western Dolomites. However, she bravely delayed our departure on hearing that our distant relative Adolf (he soon preferred to call himself Abraham) Frankel, the only mathematician related to us, was staying for his holiday at nearby St. Moritz. He had an eminent career in Pure Mathematics, becoming, relatively young, full Professor at the University of Kiel. In the early thirties he moved to Palestine, becoming Professor at the Hebrew University (and, many years later, its Rector). His religious orthodoxy was perhaps a little extreme for our taste (it is amusing to recall that as a boy I was shocked not just by the size of his family of four children, but that he publicly regretted not having more). My mother appreciated that it might become important for my future that my talent should be evaluated and noted by an internationally known figure in the field. He, equally, was intrigued to hear that there was a mathematically inclined boy in the family, and so a lengthy joint walk was engineered. Though I appreciated that in a sense I was being examined, I thoroughly enjoyed our conversation, and can still recall that in part it dealt with the distribution of prime numbers. He talked about it in a most interesting way, and was clearly impressed by how well I followed his arguments.

A very good summer holiday in Wolkenstein in the Dolomites followed, with quite a lot of walking and some energetic climbing (including the

10

Marmaloda). Returning to Vienna and school thereafter, life resumed its normal rhythm. I very much enjoyed the company of the other boys at school. I was much influenced by the excellent teaching of Dr. Mayer, a delightful and relatively young master, whose inspired teaching of history gave me a lifelong love of the subject. He also became the form master of my class, and did a great deal to ensure its cohesion. School excursions into the hills were always fun, as was the annual week the class spent skiing. This was generally in Saalbach, then a tiny and rather isolated village, now of course a major resort. Skiing in those days, except perhaps in a few fashionable resorts, was very different from today. With effectively no uphill transport you spent two or so hours climbing a mountain on your skis, with sealskins attached to prevent sliding back. Climbing with skins on is a movement I much enjoyed, but you then had only perhaps 15–20 minutes downhill skiing (not on made-up pistes of course). Though I went skiing quite a lot, on private as well as school trips, the total time (and height difference) spent going downhill was minute by modern standards. Anyway, I was quite good at skiing as at mountain climbing, sometimes to the surprise of my school fellows, as it was in contrast to my being rather poor at gym (the only official physical education in school) or at ball games outside.

Perhaps this is a moment for me to digress on a longer discussion of attitudes to sport and fitness in general. At school boys who did not succeed reasonably in academic subjects had to repeat the class and some had two such repeats in their school careers. Such boys were therefore in each class the biggest and heaviest, and so naturally excelled physically. Thus the stupidest in each class were the fittest and it was easy to come to believe that brains and brawn were anti-correlated, that the cleverest were generally fairly feeble specimens, and the stupidest were tough. This is a significant but not often discussed side effect of a policy of not allowing children to advance into the next class when their academic achievements are not good enough.

This attitude to the intellectual-physical divide was also part of the ethos of the Jewish intellectual society in which my family lived. The dull unwashed peasant lad was thought to be naturally fit and healthy, while the brainy were thought to be, almost inevitably, made muscularly feeble and perhaps hunchbacked by their many hours of bookish study. In me a dislike for this dichotomy soon arose, a dislike that my school did little to foster. Gym teachers by and large tended to be very nationalistic, inclined to the idea of "Greater Germany", often crypto-Nazis, often antisemitic, with sheer toughness their main criterion. To have to exercise bare-chested in an often cold and always dirty and sweaty gym did not inspire me to take this kind of physical education too seriously. But my climbing and my skiing enthused me sufficiently to make clear that to draw a contradistinction between intellectual and physical performance was nonsense. True, I was influenced as in so much else by my mother, whose instinctive rebellion against the received orthodoxies of our community included doubt on this dichotomy. A girl cousin was very much of my persuasion and in 1937 she and I did some stiff rock climbing together in the Dolomites.

Another little story from those days is that my parents (and my sister), while in no sense teetotallers, had little interest in, or liking for, drink. Since for social reasons the household liked to have some drink (especially of the after-dinner variety) in stock, the duty to manage this rather naturally devolved on me. I received good guidance from my elder cousin Arthur, but the sweet taste of my tender years prevailed, and liqueurs of various kinds began to figure prominently in our stock, making our apartment the favourite place for bridge parties with my school friends.

Naturally the darkening political atmosphere of the times cast a heavy cloud over this pleasant life. The total lawlessness and viciousness of Hitler's Germany, unrestrained by any normal human feelings, soon became crystal clear to me. The oppressive climate of the very unrepresentative right wing dictatorial Austrian government nauseated me and my natural youthful socialist leanings. Hatred of dictatorship extended fully to Mussolini, seen by many Jews as laudable in then protecting his client regime in Austria against the annexationist plans of Hitler and the strong Austrian Nazi movement. His Abyssinian aggression and its success horrified me. For any holidays, I greatly preferred western Austria, awash at the time with British and French tourists, and the much-admired Switzerland with its firmness against Germany. I was greatly impressed by the way British (and French) tourists and skiers combined devotion to democracy, often with a critical attitude to their governments, with total patriotism, a combination that many in Vienna could scarcely imagine.

By the early autumn of 1936 thoughts about my future began to engage me seriously. I was thoroughly self-confident in my knowledge and understanding of mathematics and especially of theoretical classical physics. I could see no outlet for my abilities in Austria. The country was a scientific backwater, with only Schroedinger in his temporary refuge at Graz a figure of international rank. An academic career in Austria was extremely difficult and unpromising for anybody, let alone a Jew. Moreover, I did not fancy doing the recently introduced compulsory service in the army, almost bound to be anti-intellectual and likely to be antisemitic, at the service of a government I so thoroughly disapproved of. Germany was obviously out of the question as a country to study in but Switzerland a serious possibility. However, England attracted me in general in a vague way. The tremendous scientific standing of Cambridge, particularly in my chosen field, was of the greatest attractiveness for me. I should say that at the time the German presumption was current that science was peculiarly a German phenomenon, a piece of nonsense absorbed by the Jewish community (regarding German Jews as German) but abhorrent to me, having learned so much of Newton, Faraday, Maxwell and Lagrange. It so happened that in early autumn the great Cambridge astronomer and writer on relativity, Sir Arthur Eddington, visited Vienna. Through a connection of my father's, I managed to meet him, and was enormously impressed by this man, splendid scientist and writer, but silent to a fault (and a very poor lecturer, as I later found out). He advised me to write to John Burnaby, Senior Tutor of his own College, Trinity, enclosing testimonials. Though Burnaby was impressed by what my school teachers had to say about my mathematics, he was not optimistic about finding a place for me amongst the small number of foreign students annually admitted to Trinity on the basis of testimonials of teachers he had never heard of. Moreover, there was no total enthusiasm in our circles for my endeavours for a number of reasons. First, because one went, as a matter of course, to the university in one's home town, and this was one of the merits of living in a large city. Secondly, because it was unusual to go to a university not situated in a major city. Third, as is true, Cambridge was so very far from home. But I immediately had the total support of my mother, and very soon that of my father as well. At my mother's instigation I wrote to Abraham Frankel in Jerusalem. He wrote to Mr. Burnaby and on his report I was readily accepted. After all, his judgement of my abilities was naturally taken at face value, since he was an internationally known mathematician who had had numerous students and so had the correct standards. This acceptance of me by Trinity College occurred about December 1936.

One hurdle remained. I had to complete my secondary school in Vienna with a good final examination (matura) in the large range of subjects involved. Of course, I had no worries at all about mathematics or physics, I could confidently expect good marks in English, German, Chemistry, Geography and History. My expected low mark in Gymnastics was irrelevant, but performing reasonably well in Latin was by no means assured. However this was needed to avoid having an exam in Cambridge which then, and for many years thereafter, required Latin. We had been taught Latin for four years. The grammar which occupied much of the first year thoroughly appealed to my mathematical outlook and I was very good at it. Similarly, in the second year, the direct logical language of Caesar gave me only enjoyment, but the convoluted writings of Livy and the involved poetry of Horace left me stone cold so that I had a problem. However, a combination of a little effort, intelligent guesswork, sheer bluff, and the respect of kind examiners for my mathematical performance won the day.

Of course I had not sat the College's scholarship examination of Trinity College, Cambridge. Thus I entered as a "Commoner", having to pay for everything. My father's income was fully sufficient, and though Austria had strict currency controls, transfer of money for study abroad was accepted. Thus no obstacles remained. After an excellent holiday in the mountains, first with my sister in Sulden where I climbed Ortler and Koenigspitze, and then some rock climbing in the Dolomites with my cousin Hedda, my father and I travelled first to Paris and then, on Sunday, 19 September 1937, we crossed the Channel to England where I have lived happily ever after.

CHAPTER 3
The first six months in England

AT DOVER we boarded the boat train. With my great interest in railways, everything about the Southern Railway fascinated me, beginning with the high platforms and consequent easy boarding which I had never seen before. Then the train, supposedly non-stop to Victoria, stopped at Orpington, where the driver, as I saw, had an earnest conversation with a senior official. The train resumed, but arrived at Cannon Street where we were met by a great aunt who lived in Golder's Green, London. Apparently there had been a derailment blocking the tracks to Victoria, and we were diverted to Cannon Street which, normally closed on Sundays, was specially reopened for our train with Customs officials and porters transferred by the bus load from Victoria. Not only did I much admire this feat of organization but I was much intrigued by a station closed on Sundays. There had been little commuting in Vienna, and certainly in my circles everybody lived well in the city. Thus the scale of London's weekday commuting was something totally novel for me.

Indeed there was a general view in intellectual, and particularly in Jewish intellectual circles, that there was little if any cultured and reasonable way of living except in a huge city of at the very least a population of a million, and that the more centrally one lived, the better. Apart from some very rich people in their villas, the members of our circles lived in flats (often quite spacious) in apartment blocks within a mile or so of the city centre. Thus commuting hardly existed, at least for professionals.

These were also my father's views, so that it was not surprising that we stayed in London as centrally as possible, at the Piccadilly Hotel. We saw relatives, admired London (and as a technically interested boy, I was most impressed by the Tube), I appreciated, as a hungry youngster, the traditional English breakfast, and then went to Cambridge to see Mr. Burnaby, still well before the beginning of term. We were both most impressed by how busy the lines out of Liverpool Street were. To see three or four other trains in motion from one's own was unheard of in Vienna. In Cambridge we saw Mr. Burnaby and the digs assigned to me (in 5 St. Clement's Gardens, Thompson's Lane), for in those days only Scholars and Exhibitioners had rooms in College during their first year. In conversation with my kindly landlady we were struck by her remark, when she complained about the poor street lighting, that she would write to her M.P. about it. To me the idea was quite strange that an M.P. dealt with his constituents' grievances and was not wholly engaged in party strife.

With our business in England done, and term still quite a few days away, my father had the idea that we might go to Paris and meet my mother there. Moreover, he thought it would be fun to fly there, having quite a few times used planes (often tiny mail planes) in Central Europe to visit patients. Thus I had my first flight, from Croydon to Le Bourget, in an Imperial Airways "Hannibal" with cloth-covered wings. The flight took well over two hours and lunch was served in the comfortable cabin. All very thrilling for a boy not yet 18! It was 1953 before I flew again.

After Paris I returned to Cambridge on my own by train and boat, and settled into my digs. Though my school English was quite good, there were obvious gaps such as my understanding of the Cambridge system and its "in" words (such as supervision, director of studies, etc.) or of technical terms in mathematics. So when my director of studies (W. R. Dean) asked each of us which was our best topic so that we could be assigned a supervisor in it, none of the terms he used was clear to me. I opted for analysis, thinking that this meant analytical geometry at which I was quite good, whereas in fact it meant the logical foundations of handling functions, of which I was totally ignorant.

I should also say that in those days one could choose between taking one's degree (always in three years) either by working through years 1, 2 and 3 of the course or, if one was very good, through years 2, 3 and 4, an option also offered. I was full of self-confidence, and told my tutor and my director of studies that I wanted the more difficult course (2, 3 and 4). Thus I found myself facing for my first supervision the formidable A. S. Besicovitch in what had been Newton's rooms in Great Court, presenting myself for second year work in a subject of the very existence of which I had been unaware. Besicovitch gave me a question. I had no idea what the terms meant. He tried me on another question, with the same result. Naturally he told me to take the easier course starting with year 1. "No", I said very firmly. He repeated his statement and I repeated mine. Evidently struck by the firmness of my refusal he said that he could only advise me to take the simpler course, but perhaps I knew better. However, he was absolutely sure that he could not supervise me owing to my ignorance, but perhaps if I learnt the topic he could teach me the following term. So I went back to W. R. Dean, who sent me to C. A. Coulson, who supervised me in applied mathematics which I found easy. However, I was utterly determined to learn analysis. Ingham's lectures were clear, as was Hardy's superb book. But it was not easy for me at first to see the point of analysis, but this soon became evident to me. Spurred by Besicovitch's dismissal of me, I worked very hard and for the first and only time in my life I suffered from a little eyestrain. My confidence rapidly grew, and in January I presented myself to Besicovitch again. He asked me to work out one question after the other and I performed well, but he remained rather silent. In the middle of the second supervision he said "Since you know all this stuff, I will tell you instead of my experiences in the Russian Revolution". This was the moment when it became clear to me that my self-confidence was wholly justified, that I had "arrived".

Quite apart from this great success of his treatment of me, Besicovitch deserves a section all to himself. An excellent mathematician, he was a teacher beyond compare. In lectures (I had lectures from him in later terms) as in supervision he constantly led his students, gently and without any malice, to pursue their arguments to an absurd conclusion, thus showing that the solution was other than they thought, thereby demonstrating the need for rigorously logical analysis as opposed to "common sense" notions. Everybody taught by him became a better mathematician as a result, nobody taught by him could ever forget his teaching or his personality. It is interesting to recall that his English at that time was still most idiosyncratic ("itinyerate" for "at any rate", a total lack of indefinite articles, etc.) We always said that you first had to learn "Bessic" English before you could follow his lectures, but Besicovitch's enormous will to communicate and to inform easily outweighed his imperfect command of the language.

While Besicovitch is the outstanding memory of my teachers of those days, I found almost all of them clear, helpful, friendly and stimulating. After having evidently overcome, at the beginning of my second term, the difficulty I first found with analysis, I was thoroughly enjoying my course, and was totally confident of my mathematical prowess. The individual attention of supervision I found far more helpful than lectures, wrestling with examples far more effective than reading textbooks, rediscovering theorems more lasting and enjoyable than taking them in from being ready served.

But I should say more about my reaction to my new country and my new surroundings. I had come to England hoping and expecting to like it, and soon my highest expectations were exceeded. The beauty and calm of Cambridge made a deep impression on me. Even more than that, the daily disproof of the faith of my Viennese circles that an educated person could only live in a large city thoroughly suited my strong anti-city views. To learn and experience from other students that life in a big city need not be the norm and was widely disliked made me feel like a fish in the water, at least in this respect.

The Viennese were intensely proud of their food and deeply contemptuous of English food, but I enjoyed it from the first. The quality of the meat, particularly when served cold which I had always liked, was infinitely superior to my Austrian experiences, while fish from the sea and especially sea food came to me as a wonderful new discovery. The light English sweets I much preferred to the heavy desserts the Austrians took so much pride in. Whereas obviously there was far less music on offer than in Vienna, virtually all my new friends enjoyed it and were reasonably informed about it, which had not been the case amongst my school friends.

Other impressions and attitudes of my early days in Cambridge may be worth mentioning. Though I had not been to university in Vienna, I knew that it was inevitably a very impersonal mass enterprise. Therefore the personal attention that one received on arriving at Cambridge made a considerable impression on me. My tutor was John Burnaby, a theologian of somewhat forbidding appearance, but the moment he started talking one became aware of his kindness and deep personal interest. Throughout my period at Trinity he was always very approachable, very interested and pleasant, but not a great conversationalist. But Burnaby was not by any means the only one whose kindness I noted. The whole attitude to students, particularly perhaps if they promised to be brilliant, showed a personal interest that was heartwarming, if a little unexpected. The general pace of life at Cambridge pleased me very much. That there were so many of us young people together, that people of different attitudes and tastes found room for their own enjoyment, all that was most agreeable. I naturally made friends, above all with the scholarly and interested, of whom quite a few will be mentioned more fully later. They were all Scholars or Exhibitioners and so lived in College. While none of us was positively poor, one didn't throw money around in our circles, but was well aware that in the College at Cambridge there were other youngsters who did precisely that. Yet others could be raucous through their particular interest in beagling or rugger, but there was very little mutual contact and certainly no hostility. Indeed this absence of personal tensions, or hostility of any kind made a considerable impression on me after the years in tense Vienna.

Of course my own attitude had also a great deal to do with this. I had disliked life in Vienna and I enjoyed it in Cambridge from the start. I did not regard myself as a refugee who, during a temporary disturbance in his own country, came to England. It was not push that brought me, but pull. It was because I wanted to be here, and I never had any doubt that this is where I wanted to live for the rest of my days. I had very firmly shaken the dust of Austria off my feet, not just because of a temporary, or so one hoped, cloud, but because by sheer liking I so much preferred England. I saw that my subject owed so much of its history to the very stones that surrounded us. To be in the College of Newton and of Maxwell was astounding. The Master of Trinity in those days indeed was that legendary figure J. J. Thompson, the discoverer of the electron, who had retired as Cavendish Professor eighteen years earlier, and was very much in his dotage by then, when he was well into his eighties. I can clearly remember seeing him shuffle across Great Court, but since I was not a scholar at that time I did not actually meet him. Not being a scholar had a more serious disadvantage for me, in that I lived outside College in 5 St. Clement's Gardens, in Thompson's Lane. Though there were a few other students looked after by the same landlady, it still meant it was more difficult and initially slower to make friends than had one lived in College. Perhaps it would also have been more difficult had I lived in College to make that remarkable burst in my studies of which I spoke earlier.

Another deep difference that I enjoyed was the importance attached to education. That the school one had been to should be something remarked on all one's life was quite novel to me and, of course, far more democratic than if it was simply one's family connections that mattered. True enough the two were not wholly different and many families sent their sons regularly to the same schools. But there were people from outside who came in, in considerable numbers, and the very idea that scholastic and educational circumstances mattered so much struck me as very progressive. Having been at a maintained school myself I took a little adjusting to the fact that many of my friends came from public schools, but this was not universal. By the time one had established oneself as one of the serious students that part of one's background mattered markedly less, if at all, as far as our life amongst our own group was concerned. Some of the minor features I can recall with some amusement, given how strict the young are about their mores. Thus, in Vienna, amongst my circle of boys, naturally one wore a raincoat if it rained, but to be seen with an umbrella was worse than death. It was a very unboyish thing to do. In England wearing a raincoat was, in my circle, rather frowned upon, but of course if it rained one had an umbrella and used it. I may say that personally I never got used to umbrellas, and my Anglicization had a relative stop at this point as also in relation to cricket and to spectator sport in general. Again in Vienna, nobody who wanted to be considered tough in any way ever wore a vest, which was considered definitely rather cissy. (This was reinforced at school since one was allowed to wear a vest in gym, but otherwise had to strip to the waist which, of course, particularly when it was a little chilly was the "right thing to do".) With the then poor heating and humid climate, vests were very common in England and for a time, if I rightly remember, I wore them too. But later I reverted to not liking them, an attitude in which I may say my two sons followed me. Similarly, in Cambridge in those days, there was a strong dislike to wearing suits. Though one always, I think, wore a tie, a sportscoat and flannels was normal. If it was a special occasion, one went into dinner jacket, but a suit was not liked and considered that it made one look like an office boy, an occupation perhaps viewed with some disdain.

Two or three of my other teachers should also be mentioned. In my first term, when as I have been saying, Besicovitch rejected me for supervision and I asked for supervision in applied mathematics, my new supervisor was a young fellow of the College, Charles Coulson who, within a few months went to Dundee and made a magnificent career as one of the founders of theoretical chemistry. Several decades later he preceded me as President of the Institute of Mathematics and its Applications.

J. A. Todd was a very fine geometer who guided me beautifully into what was for me the strange subject of geometry. I may say that the very pure projective geometry which then ruled teaching in Cambridge was something I had never met before. I knew from my friends that a great deal of geometry (essentially Euclid) was taught in English schools. It therefore came as a great relief when at my very first geometry lecture (by F. P. White, if I remember rightly), his first sentence asked us to forget all the geometry we had ever learned. This was very nice and easy for me to do as I had learnt none. With J. A. Todd's help I soon acquired quite a good proficiency in this very logical and abstract subject. J. A. Todd was a bachelor and not a Fellow of his College, Trinity. Why Trinity never elected him as Fellow is one of those peculiar questions in Cambridge. He was certainly an awkward and somewhat gawky man, with few of the social graces, but it is still surprising that it did not lead to an election to a Staff Fellowship. In his younger days I understand he missed a Junior Research Fellowship by a very narrow margin. Being a bachelor he lived outside College in a guest house, then called The Hermitage, along Silver Street, now incorporated into Darwin College, and he supervised one in his own room, which had a somewhat different atmosphere from the normal teaching room.

I soon found my circle in Trinity and became so rooted in it that before long I not only felt that my future was in England, but that it was in Cambridge. Certainly, as regards the country, I had a clear feeling that I had moved from a lower civilization to a higher one which, with all its imperfections, suited me much better and indeed inspired me. The imperfections were made only too clear to me by my friends, mostly leaning to the left and profoundly uneasy about Neville Chamberlain's "National Government" in its domestic, Imperial (especially as regards India) and foreign policies. Yet it seemed to me all the while that compared with what I had seen in Austria, the political dialogue was constructive and civilized. Both sides had positive points to make, both sides had respect for each other, yet neither side (with the exception of Winston Churchill) saw what was clear to me: the utter viciousness, the limitless brutality and the boundless ambitions of Hitler and his gang in Germany. To be truthful, while I could see no restraint on the Nazi regime's horrors, I, like others, underestimated its military strength and thought it could be contained by mere resolution and firmness.

My visit to Vienna during the Christmas vacation only strengthened my dislike for the parochial and pessimistically self-satisfied attitudes so prevalent there. At the end of the vacation I had a skiing holiday in Tyrol full of Germans drunk with the Nazi ethos, not a pleasant spectacle. In early 1938 the international scene rapidly darkened. Hitler's reconstruction of the General Staff demonstrated (or established?) his total mastery over the army. Hitler's first demands on Austria on 12 February 1938 did not seem extreme, yet the failure of Italy or anybody else to back up the Austrian Chancellor, Schuschnigg, was ominous. This was shortly followed by Anthony Eden's unexplained resignation from the Cabinet (20 February) causing consternation and gloom.

Then the pace quickened. On Wednesday, 9 March, Schuschnigg, driven into a corner, announced that there would be a referendum in Austria the following Sunday on whether independence was preferred over union with Germany. As soon as I heard this, the outcome was clear to me: Hitler could not possibly risk a, for him, negative outcome and would have to intervene by force to prevent this referendum. I therefore sent a telegram to my family of which I am still proud, that they must immediately and without further thought (*absolut unbedingt*) leave Austria. They were convinced by this urgent message from an 18-year-old boy, and on the morning of Friday, 11 March took the train to Budapest. In the afternoon Hitler marched into Austria, while my family took the night train from Budapest via Yugoslavia and Italy to Lugano in southern Switzerland where two days later, at the start of my Easter vacation, I joined them. Thus we became one of the few Viennese families none of whose members ever lived for a minute under Hitler.

CHAPTER 4
Life in Cambridge

BY THE Easter Term of 1938 I was already pretty well convinced that not only was I a good mathematician but that I should become an academic mathematician, and that Cambridge was the place where I wanted to be. All this was confirmed by the results of my first exam in June 1938 which were so good that the College immediately gave me an award, and left me in no doubt about how highly it thought of me. Altogether the warmth of the College's support and their readiness to look after me when the financial fortunes of my family suffered such a collapse owing to emigration from Austria, made a deep and lasting impression on me. Though my father's earnings had naturally ended with the flight from Vienna in March 1938, he was quite optimistic about the future. The family came to England but with their usual attitude settled in London as to them it was normal to live in the biggest city available. This choice was much against my liking. My father indeed was hoping for a while that his interests on the scientific side of medicine could lead to a research position, and was very disappointed when this did not work out. He attempted to get British qualifications, not a mean task at the age of 60, but was by then already flirting with the idea of moving to the United States. My sister, who had started on her medical studies in Vienna, went to study in Basle in Switzerland, where she completed her degree and stayed with a cousin who we all liked very much, and with whom and whose family we had indeed spent several holidays. There was quite enough money in the bank not to worry, and indeed I had good summer holidays in 1938 and 39, and a little skiing holiday in the Christmas vacation 1938/39.

Though I had long been convinced of the utter ruthlessness and unreliability of Hitler and the impossibility of making any treaty with him, the inevitability of war was difficult to understand and to comprehend. Of course I was horrified at the results of Munich, but like so many others I was relieved that war had not come. Much as I disliked the Chamberlain Government in many ways I could scarcely take aboard what was needed. After all, what was required, was not just a war like the First World War which, with all its horrors, had led to an armistice with a functioning government. But with Hitler's Germany a total eradication and destruction of governmental apparatus was required, a process that would be fiercely resisted, and so a war would have to be incredibly bloody and lengthy. So one clutched at every straw that suggested that perhaps a war might be avoided. I have always been a total realist in the sense that I believe that where a government has its troops it has control, and where it does not, what it says are idle words. When the Western allies allowed the German military occupation of the Rhineland in the spring of 1936 the pass had been sold and all the horrors that have come since had become unavoidable. I had so many friends in Cambridge by the time of Munich, that the thought of their likely fate in a war, taking of course the image of the First World War, was profoundly depressing. The re-arming of the country had my full support and I was only concerned about whether it was going fast enough. When Hitler occupied the rump of Czechoslovakia in March 1939 no hope of a peaceful solution remained. There was every impression that the Chamberlain Government too had lost any such hope, but there was the great problem of whether one could then forge an alliance with the Soviet Union. Like so many people in the West I had been rather favourably impressed by the Soviet experiment in my younger days, but with the coming of the purges and the show trials of the mid-30s that sympathy disappeared. In its place came a critical realism. In Russia there was an extraordinary system which had some elements which might not be bad, but hidden by an apparatus of dictatorship that made it very unpalatable. Yet the need to bring together a coalition against Hitler outweighed any hesitations in my mind. In these circumstances the tardiness and reluctance with which Chamberlain engaged in trying to do so aroused my suspicions, and re-awakened the idea that perhaps Stalin had been right and that the whole world was conspiring against him. With the guarantees to Poland and Romania the country was firmly set on a collision course with Hitler's Germany. The weaknesses and internal struggles in France made one already a little suspicious of her ability to give real support. Roosevelt's America certainly excited one's liking and appreciation, but what hope was there of the United States intervening actively? Undoubtedly amongst us undergraduates talk on politics was significant, listening to music was very important, but it was personal friendships that were most significant. Perhaps I may mention and describe a few of my great friends of that period now.

First, in interest and relevance must come Jack Gallagher, a most remarkable man. He came from what educationally can only be called a zero background. His father, when employed, was a railway porter in Birkenhead. They were of Irish extraction and, for whatever reason, he was an only child, a few months older than me. From what I gathered from him he must have been very advanced when he was little, and attempted to teach himself to read when he was not much over 2. Being

Irish Catholics the local priest took an interest in this very intelligent boy and helped him on. Soon, I understand, he learnt to read properly and began to acquire the voracious appetite for books that characterized his life. By the time he went to school he had already read half the world's literature. And naturally he had no contact with other children at school. Indeed, I got the impression that he had been extremely lonely until he came to Cambridge and found his intellectual class. Very political, very left inclined, yet he was a historian of great pragmatism and discernment. He also helped further to arouse my interest in the subject of history. We saw a great deal of each other and drank a great deal together. He came back to Cambridge in late 1945 and our friendship continued. Naturally it was intermitted after I left Cambridge. His brilliance in history was soon recognized. First came a Fellowship in Trinity, next a Professorship in Oxford, then return to Cambridge as Professor of Imperial History, with deep involvement in the College where he became Vice Master under Lord Butler, until his premature death. That, too, was very much in character. Not long after the war he acquired a most splendid person as his girl friend. They were a well-matched couple and we all thought very highly indeed of her. But rightly and understandably she required some commitment on his part and Jack was not a great one for taking such decisions. When it turned out that he could never make up his mind she left him, later married somebody else, had a family and a brilliant career. I do not think he ever became really emotionally involved with anybody after this period, and this loneliness made him grossly neglect his health and it was diabetes that first crippled him and later cost him his life.

Another man in my year was John Pinkerton, agreeable, practical to excess, warm and able. He went during the war into TRE and the development of radar and some time after the war went into computers, at first with Joe Lyons and then with ICL, from where he retired with considerable seniority. He married an extremely lively, able and somewhat rebellious civil servant, who was always a source of great amusement and great interest.

John Pinkerton had been at Clifton and through him I met John Kendrew, who had also been at Clifton a year senior to John Pinkerton, and whose intelligent energy and systematic application (with particular stress on the latter) had made him a legend at the school. He too became involved in radar, but very early on he went to the operational side and worked with people like Bernal and Blackett. After the war he became one of the founders of molecular biology, gaining the Nobel Prize, as befits a man of such exceptional ability.

John Jarvie was a man of particular charm whose company I much enjoyed. He came from the City of London School, read Classics, was a good all rounder but never brilliant. He joined the Navy and lost his life in the war. Early in the war he married Elizabeth, who after she was widowed, came to Cambridge to read medicine. Through her I made the first contact with my future wife, in 1946 or so. Elizabeth herself had a long career as a doctor. For a relatively short time she was married to John Kendrew. John Jarvie's sister married another friend, Ian Graham, whom I got to know through John Kendrew, very intelligent, very sceptical, who made an excellent career in the Civil Service.

Yet another friend was Stephen Ainley, perhaps my closest boon companion at one stage, a frustrated mathematician. At his school (Giggleswick) his interest in mathematics was said not to be appropriate to a boy of such intelligence, and he was driven to Classics which he studied in Cambridge without much interest or success, changing later to English. He was in the Army during the war, then had rather an interesting career in the Civil Service, including a considerable spell in Singapore. He died in 1987.

There was F. G. Friedlander, to whom I was introduced by a distant relative in London as soon as I came up to Cambridge. Also from the Continent, mathematically very bright, though he was reading engineering, he had considerable insight through his family into dissident Communist circles and their strife. He was a great help to me in finding my feet in Cambridge. He also helped me in understanding that if you are good it does not matter whether you read engineering or mathematics, or whatever.

There were many other friends in those days, so many of whom reached eminence. Jim Wilkinson, who was the best mathematician in the year above me, became very important in computer science, while A. R. G. Owen, another mathematician in my year, went into genetics with distinction. And I could go on and on. I think I have said enough to show that I enjoyed a very agreeable social life in Cambridge and that my friends were people of a wide variety of interests, but all remarkable in their way.

Perhaps I should mention one more, who had indirectly a major influence on my life, Jacques Pryce, a very agreeable young man, perhaps intellectually not quite as eminent as some of the others, but an excellent companion. He had an elder brother, also at Trinity, who was one of the Junior Research Fellows and it was through Jacques that I met Maurice. Jacques lost his life at the very end of the war in Ceylon.

When war eventually came in September 1939 one was not surprised, but I for one had certainly not thought sufficiently about what it would be like. At least I was clever enough to appreciate that the so-called phoney war from September 1939 to April 1940 was in no way phoney. Quite apart from the horrors that happened in distant eastern Europe, quite apart from the disgust at the way that the Soviet Union behaved there, in Cambridge we knew enough about the ruthless war at sea to know that this was not in any way a minor matter. It brought great suffering and great dangers. I became very interested in military matters. In a scientific (dare I say objective) way, I got my first inkling of the importance of scientific developments to warfare. At the same time I was most disappointed in the actions of the Government, or rather the lack thereof. With the war at sea fierce, the inaction and the general inability to mobilize the country were frightening. I remember the comic turn, when the Government at last began to talk of rationing (obviously necessary in a country whose importation of food was so dangerous and risky) it was said sometime around Christmas 1939 that bacon would be rationed "any minute now". Similarly the slow mobilization of forces was worrying. Finally, and more personally, though the Government were aware that there were many of us in the country who were refugees and as opposed to Hitler as anybody can be, they were rightly concerned that there might be overtly or worse, covertly, amongst us some, who by personal inclination, or through blackmail on close friends or relatives, could be unreliable during the war. The system of tribunals that was introduced to classify us gave me no grounds for comfort. I did not think that division had been as well made as I was sure that the Government had the ability to do. I had had a personal experience of this ability a year and a half earlier when after my parents had fled from Austria, I met them during my Easter vacation in 1938 in Switzerland. I read the English papers there. They were full of stories that the Nazis were smuggling their agents in under the guise of refugees. This seemed to me entirely plausible and therefore I was a little worried about what the immigration officer would think of me when I presented myself at Dover in April 1938. When I started to explain, he brushed my remarks aside, took out a book and asked me whether I had any brothers or sisters. When I answered sister, he asked whether she was older or younger, and what her name was. When I gave the correct answers he said OK you can come in. My sister had never been within hundreds of miles of these islands, and the sheer size of the intelligence effort that led to this impressed me greatly. However, I found evidence of it wholly absent during the period of tribunals in the autumn of 1939.

Just how fierce the war at sea was came home to me in a more personal manner. The London School of Economics was evacuated to Cambridge, even at that stage, and I had made a number of friends amongst them, including a very nice, intelligent girl, Lavinia Filcek (her family were of Polish extraction). Her brother was a seaman and was taken prisoner by the Germans, when one of their raiders captured his ship far away. When the Royal Navy took the Altmark in Norwegian waters in February 1940 he, with many others on this prison ship, was freed. So the reality of war became clear very readily. He distinguished himself later by shooting down a German plane with a machine gun when he was back at sea.

During the Easter vacation of 1940 I was having a little holiday in the

Lake District with some of my friends when the German invasion of Denmark and Norway happened. I remember the furious reaction of one of my friends, which was coupled with a feeling of relief that the country was at war against Hitler who did such dreadful things, linked with a feeling of powerlessness that one could do so little to give help to the beleaguered brave Norwegians. Naturally in the weeks that followed one alternated between hope and despair until the final collapse of that campaign. At last this had the essential result. Chamberlain resigned, Churchill took his place and formed a Coalition Government. My feelings of relief were indescribable. First the total confidence in the will and determination of the new Prime Minister to succeed and never to give in. Second, the end of the sham that we had a "National Government". Third, the departure from Government of virtually all those most associated with the sleepy policies it had followed. Fourth, the determination of the Labour and Liberal parties who joined it. Churchill was not the only one who went to bed that night and slept well because he knew that the country was in good hands. So that Friday was a wonderful relief and assurance of a "No Surrender Policy" in any circumstances, and thereby an assurance that, if it were at all humanly possible, of final victory. Two days later, on Sunday 12 May, I was interned.

CHAPTER 5

Internment

OF COURSE, it was not a pleasant experience to be visited on that sunny 12th of May 1940 by an agreeable policeman, asked to pack things for a few days, and then escorted to the police station where many other people had also been brought, to be transported to the barracks at Bury St. Edmunds. Our sizeable group was now a charge on the Army, who naturally were unprepared for this influx, and so to begin with things were uncomfortable, although we never ran out of food. Before long we were taken from Bury St. Edmunds to Huyton, outside Liverpool, perhaps a month later from there to the Isle of Man, where we were quartered in boarding houses, back to Liverpool and then by boat to Quebec. All the time, of course, tremendous events were unfolding on the Continent, with the overrunning of The Netherlands and Belgium, the defeat of the French Army, and Dunkirk. Newspapers arrived often, but by no means always, and every letter incoming or outgoing was censored. The reaction to internment amongst my fellow internees varied, and was considerably dependent on age. Young ones, like myself, did not find the adjustment to the uncomfortable circumstances any sort of problem, indeed found it a bit of a lark though, later, when food was not too plentiful, particularly on the Isle of Man, we felt this naturally more than older people. Older people, of course, were very disgruntled and found the whole thing a great hardship. Moreover, there were variations of opinion about what was going to happen to us. Across all the nonsense of being interned, many, like me, retained their totally positive attitude, appreciated that after the careless tribunal work of the previous autumn, there was not much that could be done at that time, other than the wholesale internment. It should be remembered that the very rapid penetration of the German Army into The Netherlands and Belgium was then reported to have been materially aided by considerable numbers of traitors, or fifth columnists. Though these stories, convenient to explain military collapse, were later found to be largely or wholly wrong, it is very understandable that the security authorities erred on the side of caution and interned us. We young Anglophile optimists were quite clear that the moment the authorities found a little time to contemplate our situation, there would be a much more careful sorting than had happened before. Not only of course did we want to get out of internment, we wanted to be useful in the war. Naturally, reactions to the advances of the German Army also differed. It certainly was frightening, it certainly was a great blow, but there was no need to be pessimistic about the war as a whole, though it was hard to keep up one's optimism in those latter days of May and first few days of June 1940.

For me, the decisive moment came when I got the news, about the 6th of June, of the Germans' southward offensive across the Somme against the French. I reasoned that Hitler's advance to Dunkirk could be due either to his strength or to Allied weakness: that the military strength of France undoubtedly had been shattered, but that of the United Kingdom had not. So if Hitler, at that decisive moment, felt unable to invade England, he was never going to be able to do so. And if he could not take the United Kingdom out of the war he could not possibly last for ever. Perhaps this optimism was not well founded, but it was there none the less. That German offensive in early June 1940 gave me the firm conviction that the war would be won. How, when and in what manner, one could not tell, but, wisely or unwisely, from that moment I had no doubts. This optimism sustained me through the troublesome periods that followed, through the many turns and twists of events.

Life in internment reverted to a rather younger period. One was in exclusively male company, and it was boyish pranks through which we amused ourselves. (I particularly recall, on the Isle of Man, any number of good-natured fights that were just an excellent form of amusement and of keeping fit.) That it was difficult to keep very clean did not bother one, again reverting to the attitudes of many years earlier. What was more worrying was how few of the Army people guarding us understood the situation at all. That so many people, officers as well as other ranks, did not appreciate what they were fighting, and that we would naturally be on their side, must be regarded as a major failure in political education. To hear remarks like, "Well the news is bad for us, but good for you, Paris has fallen", were only too common, though later things improved somewhat. It is hard to realize that so many participants in the prosecution of the war had so hazy an idea of the horrible nature of the Nazi regime, and no idea that many people had fled from it, whose very existence depended on its defeat. Even in higher circles this ignorance must have been prevalent, as is shown by the tragedy of the sinking of the *Arandora Star*, a ship full of internees, that was thought to be safe from German submarines because it flew a flag indicating that it was carrying prisoners. This sort of imagined safety meant that it was unescorted and fell easy prey to a waiting submarine. Though I am not sure how it fits in with the dates, when we were embarked at Liverpool for Quebec we took to sea similarly unescorted but, so I understand, very rapidly turned round and returned to port when the news of the *Arandora Star* sinking came. Some hours later we left with a destroyer escort, and reached Canada unscathed. It was a mid-summer crossing, July 1940, yet it was not particularly comfortable. Our first camp in Canada was on Abraham's Plains, a short distance from the landing place of Quebec. In that magnificent position, overlooking the St. Lawrence, we spent many months in quite tolerable comfort. I always said that this was the most beautiful place in which I ever lived. But this transfer to Canada also meant something else. Earlier on one thought that release would come, if not in the matter of a few weeks, then in the matter of a month or two. Transfer to Canada made it quite clear that it would be many months before we were free again. Therefore, one had to organize oneself for this, and that meant private study and studying together, for many of us were students. My first experience of teaching goes back to this camp in Canada on Abraham's Plains, and even today I occasionally meet people who say they had their best lessons in mathematics from me in Canada.

The opportunity to give lectures to groups, usually not very large, of able young people had a distinct influence on my style and method of lecturing. While I vaguely recall helping my sister when I was perhaps 10 or 11 to deal with her school mathematics, though she was four years ahead of me at school, I have no particular recollection of doing much with my friends when I was at school, largely I think because what interested me in mathematics was then miles ahead of the work we were doing at the school. Being supervised in Cambridge was a frequent and excellent challenge to explain oneself, particularly in mathematical analysis, but always to people who knew a great deal more than I did. Internment essentially gave me my first real opportunity to teach and I thoroughly enjoyed it. Lack of books (though my parents sent me some) meant that I had to work largely from the memory of what I had been taught at Cambridge a year or two years before. But this worked reasonably well. Clearly I could only teach those particular topics which were uppermost in my mind. But on those I could give my talks without notes, allowing myself to be guided by the railway tracks of the logic. This has remained my way of lecturing ever since, and in this respect internment was a formative period.

Where food and washing facilities had been scarce in England, there were plenty of both in Canada, and thus life in many ways became much more comfortable. Camp L, as it was called on the Heights of Abraham above the St. Lawrence was arranged spaciously, with separate huts, and was wide open to the wind. The authorities thought that this would be too severe and unhealthy for us in the bitter Canadian winter. Though we all liked the site, we were transferred to a much less pleasant location at Sherbrook, where we were housed in two large old railway sheds of the Canadian Pacific Railway. They afforded good protection from the weather but had none of the charm and beauty of Quebec. Interestingly enough, the fear that was uppermost in the Canadian Army's mind at the time was tuberculosis. In fact our general state of health was, and remained, very good.

But at this stage I must interrupt to talk a little about my friends. Of my circle in Cambridge, the only one who was also from the Continent and not of British nationality was Friedlander, of whom I have spoken already, and he was a great support in internment. Interestingly, being a little older, he was less cheerful than my younger friends. There I must give pride of place to Tommy Gold. We met on the first night of our internment, though we may have seen each other earlier, since his parents had known my parents in Vienna. The concrete floor of the Army barracks in Bury St. Edmunds was an odd place to start a friendship that was of greater influence on my scientific development and, to some extent, personal attitudes than any other of my generation except Fred Hoyle. At that stage Tommy was a bit of a playboy. Though only six months younger than myself, he was two years behind me and read engineering, also at Trinity. His intelligence, positive outlook on the world, and wide technical interests were immediately apparent. The depth of his thinking and his thoroughness took me a little longer to discover. He was no mean athlete in a number of fields, and an exceptionally fine skier. Perhaps this was not surprising in somebody who had been in school in Zuoz in Switzerland, with wonderful opportunities, but his whole posture and physical abilities were remarkable, whether it was a question of walking on a rope or skipping.

In retrospect, it is not easy for me to distinguish when I first began appreciating this or that feature of Tommy's mind, for our lives were generally shared from this first meeting in spring 1940, for over a dozen years. He certainly made it clear to me, in a way nobody else had succeeded before, that intelligence was not particulate; that if I was good at differential equations then I should be equally good at understanding the engineering problems of a motor car engine, or the design of a radio set; that thinking about economics was no more reserved for particular turns of mind than discussions on warfare, a matter in which I had long been interested.

I do not think he ever began to share my liking for history, but there were any number of topics on which we were mentally close. Even nowadays, half a century on, we can discuss the most serious scientific topic within a few minutes of getting together. In many ways much more thorough than I, for example in reading (it may be amusing to recall that years later, reading Agatha Christie's *The Murder of Sir Roger Ackroyd*, he spotted the significance of the doctor's phrase "I did what was necessary", as soon as he read it, so that the outcome of this detective masterpiece was obvious to him from the start). Of course this thoroughness sometimes meant a certain slowness which, though irritating, was always a good corrective to my shoddiness. Years later, when we wrote scientific papers, I always had to keep a check on his perfectionism, which could delay publication considerably. His sound temperament, a great readiness to help me in the practical problems which I was ill-equipped to handle, a common enjoyment of seeing the funny side of life, all these made us inseparable friends, illuminated a large part of my year in internment, and inspired me many years later in my career. A fourth one in our group was a man called Martin Silver who, before too long, dropped out, but who was also a good companion at the time.

Of course, one read the news greedily, of course one eagerly waited for the release that would allow one to return to a normal form of life. Of course, one was only too anxious to be able, one way or another, to be of some use to the war effort, since victory was so essential to life itself. Yet with all the irritations of being a prisoner, I do remember particularly that the early autumn in Camp L was really rather an agreeable period. I did attempt to do some scientific work of my own, aiming for a Trinity Fellowship, a matter close to my mind since in October 1940, Friedlander was in fact elected a Fellow of the College.

I now need to return to the rest of my family, since there were important interactions to come. My sister was already clear, fairly early on, that she was unlikely to be allowed to practise as a Doctor in the United Kingdom, and so she applied for an American immigration visa, determined to go there as soon as she could (waiting lists were of course quite long at the time). My parents similarly applied but more as a fallback position than anything else, because they too rather liked it in the United Kingdom. My sister, in fact, entered the United States from England, to which she had returned from Switzerland in the early summer of 1939, having completed her studies there. She went across to America in the early spring of 1940. In the early summer of 1940 my parents, too, decided to move to the States and in the early autumn they managed to do so. Like so many non-British people they despaired of the war situation after Hitler's victories on the Continent in May and June of 1940. The mood in much of the world was the same, with the United Kingdom's chances of survival viewed as slim or non-existent. A respected U.S. commentator said: "I am an optimist. I regard the future as uncertain." So my parents were not alone in their views. Many years later a Swedish diplomat involved in getting permits from the combatant nations and therefore a frequent visitor to both London and Berlin, told me of the contrasting moods of the two nations. In London, on the underground, the general attitude was "Sure, this is a rough spot but of course we will win in the end." In Berlin, while the younger generation was drunk with the victories gained, the older people winked at each other and reminisced about the great successes of the German Army in spring 1918, which were followed by collapse some months later. In particular it was worries about my fate, as a helpless internee, liable to be handed over to the Germans in a British surrender, that they saw as a compelling reason to leave the country much as they liked it. For once they were in the United States, they thought they could get me, still a minor under the laws of the time, the necessary immigration visa to enable me to join them there. They had had a most anxious time in England about me, when the news came of the sinking of the *Arandora Star* with the heavy attendant loss of life amongst internees, before they could find out that I had not actually been on board that ship. With the pessimism about the chances of England current in the New York circles which they joined, with the horror at my internment (much greater than my own), they naturally saw my future to be in the United States; while I had the opposite intention, namely returning naturally to the United Kingdom, where my College valued my abilities, where my friends lived, where I felt utterly at home, and where I thought that I would be much better able to help the war effort, than in the United States, then still determinedly neutral. Of course, as I explained before, I was also totally convinced that we were going to win the war, and that the United Kingdom was not at risk.

This was the situation when the news came to us that the government of the United Kingdom had appointed a distinguished civil servant, Sir Alexander Paterson, well known for his broad and humanitarian views, to sort out the internees and arrange the release of all those where he felt this could be safely done. But of course this took him a long time, and those who were in Canada were certainly not at the top of his list of priorities. My parents, with my cousin Arthur as a lawyer already established for some time in New York, meanwhile got busy, arranged a visa for me and in I think January 1941 (but it may have been in December 1940) I was taken out of my internment camp to the American Consulate in Montreal, where I got my visa, but was immediately told that this did not settle my entry to the United States, because U.S. Immigration also had to be satisfied. By the regulations then current a person coming from other continents could not enter the United States from a contiguous country (this meant Canada and Mexico) unless he had paid his fare to those countries. Of course I had not paid my fare, and so while I had obtained a U.S. visa, my immigration was denied. It is difficult to express my feelings on this. On the one hand I was really very set on my future in England, on the other, freedom beckoned across a border a few miles away, and I was getting a bit tired of being behind barbed wire after eight months or so. Finally, my parents' letters told me of the serious heart attack that my father had had in the autumn of 1940, so my feelings were mixed. The stupid legalism involved in my being refused immigration turned me further against the United States, as did the reports of the attitudes in America to bringing in internees where again many people did not understand the situation too well, and one was not at all certain of a friendly reception, let alone one that would recognize my abilities speedily, if at all, so I could put them to good use. Especially when, as we all hoped, the United States would eventually join in the war, I wanted to use my mathematics to defeat Hitler. I was then put in an internment camp on St. Helen's Island in the St. Lawrence River at Montreal, so that I was available should the appeal against the refusal of my immigration be successful.

There follows a rather unhappy period when the likelihood of my release, due to British initiatives, increased, when my desire to go to the United States faded, when I was still interned in not particularly pleasant conditions, with much of my company Italian sailors interned in Canada, but initially at least my family pressed rather hard. Then came the time of the interviews with Sir Alexander Paterson and his staff, and the return to England was beckoning. My parents, particularly my mother, had by this time totally accepted my return to England. But a more distant relative took it upon himself, in a most irresponsible manner, to contact Sir Alexander Paterson himself, and tell him it would be very wrong for me to be returned to England, given that my parents and especially my not very fit father were hoping for me to join them in the States. Paterson, very understandably, said that he would leave it to me and my family to sort things out. But the ill-judged intervention of this relative cost me three months more internment.

Much of these latter phases were not as agreeable as the earlier ones. The company was quite different, generally ill-educated, so there was none of the teaching I had enjoyed. Also the Canadian winter, in the middle of the frozen St. Lawrence, was nothing like as pleasant as the summer above St. Lawrence, near Quebec. Moreover, my impatience was great. Eventually the great day came and I, and some of my lucky companions (none of whom was a close friend), were taken on the long train journey to Halifax, and then sailed in a Canadian troop convoy to Scotland. This was in June 1941. Not long before the crossing of the Atlantic, the *Bismarck* episode had occurred, which certainly would have been very worrying had one been in mid-Atlantic at the time. While I was at sea, the German invasion of the Soviet Union took place, ending the long period when the United Kingdom with the rest of the Empire stood alone against Hitler.

In early July 1941, we arrived at Greenock, but I was still in internment and transferred back to the Isle of Man, and my final release did not come until the beginning of August 1941, when I was landed at Fleetwood with a ticket to Cambridge in my pocket. In the middle of the long vacation I was back at my home of Trinity College, from which I had been taken fifteen months earlier.

CHAPTER 6

Interregnum

MY STATUS in Cambridge was clear. While interned the College had arranged for me to receive my B.A. degree. The Part Two Mathematics Tripos that I had taken in the early summer of 1939 was, in fact, the degree-giving examination, but a degree could not be awarded unless one had three years residence as well. In the academic year 1939/40 I had been reading for Part Three of the Tripos. This was enjoyable as a preparation for research rather than as a necessity for the degree. My internment in May 1940 of course meant that I had not strictly fulfilled all the requirements of residence, but this was readily excused. My senior scholarship had automatically been turned into a research scholarship and so when I arrived back in Cambridge, admittedly in the middle of the long vacation, it was clear that I would be a research student, and I registered for a Ph.D. The centre of my interest had been in Classical Applied Mathematics, particularly the theory of waves, and so it was not difficult to find who I would like best as my registered supervisor, namely Harold Jeffreys, a great mathematician and geophysicist, active in many fields of research including, particularly, waves. He had no doubt that I was ready for research and immediately gave me a small problem connected with waves on the surface of the water, and within a few months I had written my first paper, not a particularly distinguished effort, but readily publishable in the *Proceedings of the Cambridge Philosophic Society*.

I found the work interesting and absorbing, yet I was much perturbed by not being able to do anything to further the war effort. I turned to Maurice Pryce, who I have mentioned before, a Junior Research Fellow at the time, who promised to use his contacts. He was only very occasionally in Cambridge, being already deeply involved with the Admiralty, and before very long he told me that he was very hopeful of finding something suitable for me without, in those days of secrecy, telling me, or my even expecting to know what this would be. The paper work and getting all the necessary permits, something with which I personally was not concerned, seems to have been a lengthy and tiresome matter, and only at the very beginning of March did I receive the instruction to report for work at the Barracks of the Royal Marines, at Southsea, near Port-

35

smouth. It was left to me, if I rightly remember, to arrange the precise date which was in fact the first of April 1942. In connection with my later academic career it is amusing to note that the total time I spent as a research student was the seven months from September 1941 to the end of March 1942, at which time I started full-time work in my first paid employment.

Thus while my scientific and employment history of the interregnum is rapidly told, its personal aspects were rather more important. My friend from internment, Tommy Gold had been released three months before me, and as soon as I was released I made contact with him. This not only meant that I had splendid company in Cambridge but, more than that, I had his circle of friends as well as mine (he had made contact with many of my friends as soon as he returned to Cambridge, though of course many of them were by that time in the armed services). I also met his friends and, particularly, his parents, who gave me a "home from home" where they lived in Highgate, and I had a great deal of kindness and pleasure from them. I saw much of John Cox, an old friend of Tommy's, of Lavinia, who I mentioned before who, in the meantime, had become Tommy's girl friend, of two members of her circle at LSE, Stephen Wheatcroft and Joy Reed, and many others. Tommy and John Cox, of course were still undergraduates, and I reverted readily to my role of teacher. Tommy's deep physical insight and technical interest were not at that stage joined to particularly good mathematical competence, which gave him a lot of trouble, and so I could be helpful. It is an interesting reflection on the teaching he must have received, that he claims to this day that it was from me that he learned the importance of dimensions. Several times he handed me the results of what had been for him a lengthy calculation, I glanced at it and said "That's wrong." He was naturally furious that I could dismiss, at a glance, what had been a lot of effort for him. When I explained that the dimensions of his result were inconsistent, he was duly impressed and found this very instructive. Tommy, John and I were very close companions, and spent a great deal of time together. I acquired the gift, lost much later with advancing years, of being able to concentrate absolutely and completely on my work while other people were in the same room, chatting away to each other, and occasionally involving me in their conversation. All this was played out against the background of the drama of the war, of occasional messages from our friends in the forces and, even more occasionally, personal appearances of theirs; of growing shortages of food, of difficulties of transport, of the now only sporadic bombing. The progress of the war was watched naturally with the greatest excitement and involvement. After the rather poor performances of the Soviet forces in the winter war with Finland, the rapid German advances of the summer and early autumn of 1941 did not come as a surprise to us, but we were greatly heartened by the stiffening of the resistance before Moscow, and the German failure to take Moscow in the autumn of 1941. The British autumn offensive in the North African desert was only a partial success, which was a great disappointment, but the growing strength of the RAF was a constant source of encouragement.

Into all this came the Japanese Declaration of War with their attack on Pearl Harbour in early December 1941, followed within a few days by the sinking of HMS *Repulse* and HMS *Prince of Wales* off Malaya. Undoubtedly this event was one of the most shocking of the war. I think one had a fair idea that together with the American disaster at Pearl Harbour, this meant that Japan now had a free run over a large area. We had the greatest worry about the ability of Australia and New Zealand to escape Japanese occupation: two countries, very close to our hearts because of their consistent help during the war, their civilized social structure, their hoping against hope in the first few weeks of 1942 that the fortress of Singapore would prevent further Japanese expansion and would be a safeguard to them. In all this collapse we were greatly heartened by the prolonged American stand in the Bataan Peninsula, and the charismatic aura of General Douglas McArthur originated in those days. The miserable episode of the escape of the German warships through the Channel in mid-February 1942 added to the general gloom. Yet even in those days I had no doubt of the eventual outcome of the war, but it was impossible to see how this would be achieved. Of course we placed great faith in the United States, but also in British strength, and the persistent success in preventing the fall of Malta in the Mediterranean, coupled with the build-up of the RAF gave us great heart.

I still remember the impression I got when I saw the first four-engined bomber, the Stirling, in the air. Here was an achievement of the British aircraft industry that boded ill for Germany. The bomber offensive clearly was very much in our minds and a great boost to morale. Though compared with what was to come later those early months of 1942 were only a modest beginning, yet we could see that the air superiority had definitely passed to the United Kingdom. Here was the one way in which the United Kingdom could hit Germany and hit where it hurt most. To the modern reader it may sound strange that none of us had the slightest qualms about employing bombers against civil populations. Hitler's Germany had to be destroyed, otherwise there was no future for mankind. Its strength was based on its military industrial machine supported, so it was clear, by a population which, with few exceptions, followed Hitler during his successes. There was no sympathy of any kind for this population, but there was a distinct worry about whether the bombers could be employed to the best effect in wrecking this industrial capacity. As we all know, this became later a major bone of contention.

CHAPTER 7
Admiralty Signals Establishment

IT WAS at Portsmouth, or rather Southsea, on the 1st of April 1942 that I first was gainfully employed as the quaint official terminology calls it. Until that time scholarships, academic support and my parents had backed me. Now it was from the fruit of my own labours that I lived. I discovered that I was to work on radar, or rather RDF, as it was called at the time. The establishment was a branch of the Admiralty Signals Establishment, ASE, who now counted me as one of their Temporary Experimental Officers. Though I had vaguely heard of radar it was a revelation to me to discover the power and strength of it, the ability to detect small targets at great distances, and the prospects of yet higher performance.

I had joined Maurice Pryce at the barracks of the Royal Marines. It was a strange experience for me. As I have stressed throughout, my work had been very theoretical. I was purely a pencil and paper man, and now I was thrown into a research establishment that was attempting to design better radars for the Royal Navy, radars of a kind that could be kept in good working order, not just by white-coated lab technicians with nothing else to worry about, but by sailors in the middle of a battle. This was an enormous adjustment for me, but perhaps the biggest one was to appreciate that my engineering colleagues had such a strong instinct for how things should work, and such simple yet effective images. I knew Maxwell's equations backward and it was their application in which I had some experience. Such application should, I thought, give one a perfect answer. Until I became immersed in the work of ASE I always thought that one either knew the answer to a problem or one did not. No doubt this was the result of the examination system in which I had been brought up. But it is also very much in the nature of the work I had been doing until that time. Now I had many colleagues who made things work without fully understanding them; colleagues who seemed to have an instinctive feel about matters electrodynamic. So my whole way of thinking had to change. I lived with a landlady in Southsea, who looked after me very kindly, and work occupied me almost wholly. Portsmouth was occasionally bombed, and I remember being blasted out of my bed one night. The blackout, of course, was intense. Rationing was in full swing.

Yet I have very pleasant memories of that time. While job-wise I was essentially learning and had no great feeling that I was a real contributor at the time, I very much enjoyed the atmosphere of Portsmouth (incidentally, it is the biggest city I have ever lived in since I left Vienna). Almost all young men were in the Services, but when they saw civilians like me and my colleagues there was not the slightest antagonism. There was total confidence in the authorities that they had asked us to do a job we could do better than anybody else, and that we were all doing the same thing, working for the same aim. More than that, the camaraderie was astonishing. Once or twice I remember hearing in pub-talk items which I knew from my work were supposed to be very secret, but I also knew from my work that they never got across to the enemy.

I stayed in Portsmouth until the early summer of the year. While most of the work of the establishment was carried out in the Marine Barracks at Southsea, this was not a suitable site for work on aerials for which there was a branch establishment at Nutbourne, near Chichester, maybe twenty miles away. I heard then that there was an interesting "wild" mathematician there, Fred Hoyle. I think I must have met him once or twice during those three months, because he had a certain amount of contact with Maurice Pryce. In the summer, we were suddenly moved to King Edward School, in Witley, in Surrey, in the beautiful Surrey countryside that I had not known before and for which I then developed a love which lasts to this day. At the time we all thought that this move was a very belated bureaucratic response to the heavy bombing of Portsmouth a year earlier. It was much, much later that I learned what it had all been about: that spring our Commandos had carried out a very successful raid at Bruneval on the latest German radar, and brought it home virtually intact. The powers-that-be rightly thought that what could be done one way round could be done the other way round. Therefore the new equipment that was being developed should not be sited on the coast, even in Marine barracks. And so we came inland. The poor school apparently was expelled with practically no notice, though rumours of the move had been current amongst us before.

Not long after our arrival there was a complete reorganization. People from Nutbourne and the previous Southsea establishment were brought together. A Theory Division was formed with Fred Hoyle at its head, and myself as his deputy. Originally it was very small and I think consisted only of two other members, an Oxford mathematician, called Gillams, and Cyril Domb, also ex-Cambridge. We were joined by E. T. Goodwin who, after the war, achieved considerable note in working on computers at the National Physical Laboratory. I believe I am right in saying that no more than seven or eight people can claim ever to have belonged to this little group, but five of us became Fellows of the Royal Society.

Histories of radar have been written by others, and I do not want to go into detail, but I do want to mention the outstanding head of all this group of divisions, a man called Landale, who had been Managing Director of Youngers Brewery and was a most forceful and effective leader. In particular, he had the great fame of having got the first really short-wave equipment (at 10 cm) into the hands of the Royal Navy far faster than anything had been done before. This was a life saver in anti-submarine warfare and he had a tremendous standing amongst us.

We were billeted amongst large houses in this well-to-do part of Surrey, and I remember staying with very pleasant people called Palmer who really were most helpful in making me comfortable in not always easy conditions, and who strongly felt that in supporting me they were supporting the war effort. The respect for the scientific side of it all that was so widespread, was really very remarkable. My friendship with Tommy Gold was such that on many weekends (of course we worked on Saturday, so the weekend only started late on Saturday afternoon) I went to London both from Southsea and from Witley to stay with Tommy Gold's parents, where he too spent many weekends. In the summer of 1942 he at last got his degree. Maurice Pryce, Fred Hoyle and I, on my recommendation as to his gifts, tried to bring him into the business of radar, into our group, but his poor degree combined with the natural slowness of the machine meant that it was late November before he joined us. He was billeted in the same house as I and joined our Theory group. In the meantime he had done his bit for the war by working on a farm in the Southern Lake District, which was then Lancashire.

It is worth recalling from those days that although all long distance calls had to be made through operators, the phone system worked superbly and one was put through very rapidly. I called him from a remote corner of Surrey to a remote corner of Lancashire (the exchange, if I remember rightly, was called Lowick Bridge). I always got through speedily, and so he could follow the progress of our efforts to bring him to join us. In retrospect I find it remarkable how much confidence was placed in the judgement of a young and inexperienced person like me, who had been in the business of radar only for a few months. So I managed to mobilize all this support for Tommy joining us. Although he came from a different subject and had not then the outstanding academic support that I had had, yet it all worked well. Tommy's practical bent did not enjoy being billeted on people however kind and helpful they were; he wanted independence. In this as in all practical matters I was glad to be guided by him. We looked for a little house of our own and, before very long, found one in the delightful village of Dunsfold. It was rented out by a farmer called Renmant and his charming wife, who came from the French-speaking part of Switzerland. They had built themselves a new farmhouse, then rented out the old one which was to be my home for two and a half years, from the very beginning of 1943 to my return to Cambridge in the summer of 1945. I have the happiest memories of staying there, through my friendship with Tommy and through Fred Hoyle. He, who was four years older than I, was already married with a child (and a second one on the way before very long) and they had a house near Nutbourne, while Fred was now stationed at Witley. Rather than go through the trauma of selling and buying houses he spent much of the week with us and commuted weekly. In a three-bedroomed house it was not difficult for Fred to have one bedroom, Tommy another and me the third. I acquired my first taste there for domestic engineering and for cooking, under Tommy's guidance, but he and I and Fred, when he was with us, spent all our time discussing scientific questions. Fred's enormously stimulating mind, his deep physical intuition, his knowledge of the most interesting problems in astronomy, all combined to give me an outstanding scientific education in the few hours left after a hard day's work.

Life at Dunsfold was very agreeable though, of course, with the long working days, we did not spend that much time there. It was a truly rural setting, something that I have liked and desired ever since. It was very quiet, at least until about a year and a half after we came there, when an airfield was built near Dunsfold for fighter bombers. They took off over the little rise where our house stood, usually very heavily laden. Hearing those planes pass us over the rooftops could be a bit nerve-wracking, particularly as we knew that occasionally they crashed, blowing up their bombs, as had happened not very far from us on one occasion. In this remote location, we had far less to fear from enemy bombers than from British ones.

Under Tommy's guidance we organized our lives well. We got a cleaning lady to "do" for us, who cycled over from the next village of Cranleigh; of course she arrived after we had left, and left before we had come back, so communication was entirely written. But she very much realized what we needed in the way of cleaning and washing and clearing up the house, and was most helpful. Indeed the whole village was rather sorry for these two young men who had to manage for themselves, and this attitude had considerable advantages for us. Rationing was severe at the time and our appetites were as good as one can expect from young men of our age. Most foods were rationed but some special ones, like chickens were not. They were just difficult to get in general, but in the village a local farmer saw to it that we had one for our lunch every Sunday. The farmer and his wife were equally kind to us. Every now and then Mrs. Renmant would knock on the door with a bowl and said, "We killed a pig, would you like some brawn?" which, of course, was always most welcome to us. However, being surrounded by the farm also had its disadvantages, for every now and then the pigs, or other animals, broke through into our little garden. I remember on one of those occasions Fred, saying, "Should we now go across and tell Mrs. Renmant 'We killed a pig, would you like some brawn'?" The house was adequately if sparsely furnished with space heating entirely by the fire in the living room, and its back boiler heating the water. Our cleaning lady left a fire laid and we lit it every evening, because, of course, coal too was not over plentiful. Nor had the technology of keeping fires in all day spread to our little cottage. I remember coming home one cold evening with Tommy, and I went to light the fire and it went out. I rebuilt the fire, it went out. Tommy rebuilt it again, it went out. So we said, "Oh it's pretty late anyway, so don't let's bother ", but since it was pretty chilly walked up and down in our overcoats discussing, as usual, some scientific problem. Tommy, who was then a smoker, lit a cigarette, threw the match carelessly into the fireplace which, in a moment, was a blaze of flames, and the fire we had so vainly tried to light by intent was lit accidentally with perfect success.

One of the problems of being in that house, though, was its remoteness. One could cycle to work, but it was quite a good distance and rather hilly. Though we cycled gladly in fine weather, we were not so keen when it was wet. There was public transport from Dunsfold to one end of Godalming, then we had to walk through the little town and catch a train at the other end to Witley Station which was very near our establishment. However, catching the first bus (at 8.25, if I rightly remember) still meant that we did not get to work until about 9.25. Tommy then bought a car, an old Hillman. Petrol was severely rationed in those days. There was no basic allowance, one could get fuel only for purposes deemed essential. When we applied for petrol on the basis that we could not get to work at the required time by public transport, we were given coupons for all the petrol we asked for, and a little over, and for all the other things that were rationed, such as tyres. Being young men, not too keen to get up early, it was rare after we started using the car that we turned up at the office before a quarter to ten or so. However we worked until late in the day, so that we grew a little tired. We had usually a good evening out on Saturday night and slept quite late on Sunday. However, we firmly insisted that on Sunday one had to have breakfast, lunch, tea and dinner, finishing in time for a last drink at the local pub, which closed at 10 o'clock. Since we often did not rise until about noon that meant that the Sundays were very strenuous and full indeed. The oven, of course, was also coal fired with no automatic means of getting it going, which made the timetable on Sundays rather difficult. Tommy, ingenious as always, then constructed an automatic way of lighting the oven. The alarm clock, the bane of our lives during the other six days of the week was moved into a position where a spanner was balanced on it. When it started to ring, well away from our bedrooms (in fact, we deadened the sound with a little bit of paper), it knocked off the spanner, which was attached to an electric switch, the wire shorted through a fuse wire woven into a piece of film which led to the fire that had been laid on top of a fire lighter. More often than not, this Heath Robinson piece of machinery actually worked, and we woke mid-day on Sunday to the splendid smell of roasting chicken. Work, the purpose of all we did, consisted of two different parts, each challenging, each interesting, each fascinating. One was our official work, which made up our duties and for which we were paid meagre salaries. (I was a temporary Experimental Officer, with a salary of £300 a year, later raised to the princely sum of £350 a year.) The other work consisted of our energetic discussions of scientific problems, largely taking place in the evenings after our work for the Admiralty, and centred on astronomical questions. This often led me to work mathematically late into the night. As regards the official work, the Theory Group had many duties. Perhaps our prime duty was to handle queries as they came to us. One major area was to study the performance of existing equipment and forecast from the design what the performance of future equipment would be like. To handle this I had to rediscover and adapt the theory of noise which was the limitation on the performance of the radar sets. The problems of fading were more often dealt with by Fred. The possibilities and limitations of aerials were again a very major subject, where we worked closely with a great aerial designer, also a refugee from Germany who had been very eminent there, a man by the name of Boehm. He had a wonderful intuitive understanding, and from him I first learnt how, by experience and knowledge he could see in two minutes intricacies which took me, with all my mathematical skill, a week to re-derive. Yet the work I did was a significant reinforcement of his efforts. The understanding of the way wave guides worked was then rather limited. And again we managed to improve understanding, and so assist with the engineering. A very important subject for the Royal Navy, on which a lot of work was done, concerned the propagation of centimetric waves above the ocean. The gradient in water vapour density had remarkable effects which could lead to very long, indeed disadvantageously long ranges at times, while at others the phenomenon was absent. The phenomenon was already important operationally in the case of 10 cm waves, for 3 cm waves, to which one hoped to progress, it could be very serious indeed, and the refraction could hide targets as much as enhance range.

There was no easy way to carry out continuous trials, for aircraft for trial purposes were hard to obtain, and could be used only intermittently, whereas what one wanted was to know what the record of these propagation variations would be over a long time, and to examine their correlation with the weather. It was, I think, Fred Hoyle, who had the idea of using the summit of Snowdon in North Wales, easily reached by rack and pinion railway, for this purpose. At an altitude of nearly 1100 metres it was not much lower than the standard height of aircraft flying in those days. The sea was not many miles from the peak in many directions, and the idea then was to study propagation between the summit of Snowdon and the Experimental Establishment at Aberporth, on Cardigan Bay, in Southern Central Wales. This would give a path over the sea for most of its length, above Cardigan Bay. Then the question arose who should run this equipment on Snowdon. Although, as I have stressed, I had been very theoretical in all my career, the attraction of the hills was just too great for me. I knew the area a little bit from a holiday spent with John Cox, in Nant Gwynant, and found it most attractive.

Anomalous propagation naturally occurred mostly in the summer, so the original plan was that the Station, recording 3 cm and 10 cm links between Aberporth and the summit of Snowdon, would be working for the summer of 1944. To get the whole thing set up involved our administration in great problems, but eventually it was all done: the café on top was taken over by the Admiralty for me, my staff and visitors. The Snowdon Railway was commandeered to run solely for us, the top was suitably surrounded with a little barbed wire to prevent unauthorized ingress, and all looked fine when I first arrived on top, except that all our administrators had been able to dig up to power the job (which they never liked anyway) was an old paraffin generator. Although it had just been decarbonized it very rapidly died on me, and the whole business would have been wrecked had Fred not used his link with Sir Edward Appleton to get a hold of two splendid Lister diesel engines from the army. They were my companions for all my stay on Snowdon, and their reliability was outstanding.

Though the original plan for Snowdon had only been this propagation experiment, another issue soon came to the fore. Radar was used from aircraft to find submarines and one of the problems, particular when only the periscope was showing, was the return from the waves of the sea, particularly for the higher frequencies that were needed for such low targets as submarine periscopes. The Americans had just experimented with using a whole sizeable aircraft to house a big radar and look down on the sea from it. They claimed to have had good success in their proving ground of Boston Bay, a notoriously calm area of the sea. We were extremely doubtful whether this heavy investment was worth while in the wild waters surrounding the United Kingdom. Therefore, while I was busy on Snowdon with the propagation experiment, a requirement arose for studying wave clutter, ideally with targets of known size against it, but in any case to get an idea of how wave clutter on the radar screen diminished with distance and varied with sea state. Of course Snowdon, surrounded by sea on three sides, was ideal for this. But, whereas for the propagation experiment summer was ideal, because then the water vapour concentration above the ocean was at its highest; for the wave clutter experiment, one preferred the rough parts of the year, autumn and winter. So what was intended to be a summer job turned out to be a much longer one, taking me far into winter. Moreover, in addition to the relatively modest sized propagation experiment equipment a big radar, type 277, operating at 10 cms, had now to be brought up to the top and installed. Fortunately the big diesels that Fred had got for me were up to the task, but the overall logistics became distinctly tricky as winter approached. The Snowdon Mountain Railway, a rack and pinion line dating from the 1880s had had a nasty accident very early in its life, an accident which, rightly or wrongly, was ascribed to a tough stone having lodged in the rack, raising the engine's pinion wheels off it, so that it slipped back and derailed. The firm policy of the Snowdon Mountain Railway was that, with an engine pushing at the back, as is normal with rack and pinion railways, there always had to be somebody sitting right in front to watch the rack, to make certain that there was nothing lodged in it. Evidently in winter, when there was snow, there was no possibility of doing this, and so the railway could not run in winter, or in the dark. In summer we used the train for passenger and goods transport, but the Company thought it quite sufficient just to have the open goods truck for us as well as for the equipment. On a wet, windy day, spending that hour on the truck was not necessarily very comfortable. For the summer I had been given as my staff two civilians, a soldier who had been invalided out of the Army after injury, and an elderly man. They were extremely helpful but hopelessly non-technical, and everything technical, be it the diesels, be it the wiring, be it the heating, be it the experiment itself, had to be done by me. Of course I could ask for assistance, but inevitably that assistance took some time to come, so I learned to do a great deal myself. I loved the mountain, and walking instead of taking the train was not uncongenial to me. When the weather was good, which is not often on Snowdon, and I could spare a little time from my work, which indeed was very absorbing, I used to clamber about. Also Fred came to visit me occasionally, as did Tommy and other companions from my group and elsewhere, usually very briefly.

Bad weather could be irritating in summer, severe in winter. In summer I remember a period of over two weeks when every day there was very little visibility, gloom and often rain or sleet. Yet when I went out of the compound to talk to climbers who had walked up from below, they invariably said "It's a lovely day, there is just a little cloud round the top of Snowdon." Since I was too busy then to get away this was little consolation to me. Yet, on another occasion, there was a truly glorious day, and I just could not resist the temptation to run down the east side of the mountain and come back which would not have taken too much time. In that lovely summer weather I only wore boots and shorts and of course took no food with me. When I had gone some way, I found that it was such a singularly fine day that I simply had to get a little further and do the whole horseshoe of ridges emanating from Snowdon. I was very fit so it did not take me all that much time, but on the sharp ridge of Crib Goch

I was beginning to feel hunger pains. Rounding a corner, I came across three disconsolate boys. They said they had lost their way and didn't know what to do. I told them I would happily act as their guide, and give them some food when we came to my Station on top of Snowdon, but could I please have some of their sandwiches then and there. It turned out they were all undergraduates from my College, Trinity, very young as students were during the war, including one very bright looking one, J. C. Sheperdson, who in all his younger days (and I saw much more of him later) always looked inevitably two years younger than he was, being slight and slim. He became a very eminent mathematician and Professor at the University of Bristol.

With Fred (I did not drive myself in those days) on various occasions we went down to Aberporth to have the necessary discussions. The Aberporth end, incidentally, was run by E. T. Goodwin, a member of our Theory Group I have already mentioned. On the way back Fred invariably misjudged the time it would take. The Railway, of course, would not run up so late that they could not get down in full daylight, because of the worry about the rack. And so we frequently found ourselves far too late for the train, in which case we usually stopped at Penygwridd and walked up in the pitch dark along the Pyg track. One of those excursions was perhaps a little more dramatic. On our return drive we crossed in the evening the pass that leads down to the Cross Foxes Inn. Suddenly there was a dragging sound from the back. Fred immediately stopped and jumped out. What had happened to this little naval Morris was that the petrol tank had become detached and was held only by the tube leading to the carburettor. Of course if the car had gone on and dragged the petrol tank, it would soon have become holed which would not only have been very dangerous, but would certainly not have allowed us to continue. As we stood on the broad empty road in that vast landscape, wondering what to do next, we saw that the farmer who had fenced the field next to the road had left lying around a good few of the offcuts of wire. Fortunately it was not barbed wire. With a good deal of muscle power we attached the petrol tank to the chassis with those wires. We did not trust this repair at all and every quarter of a mile Fred stopped and I jumped out to look whether it was still holding. It always looked perfect, so we lengthened the interval to a mile, then to five miles, then we drove through the night to Penygwridd and walked up Snowdon. By that time we had become so convinced that the attachment was all right that Fred ran down from the top next day to Penygwridd, took the car and drove it all the way to Edinburgh. Indeed our repair held and when he took it to the garage to ask for the fixing to be restored to the more normal one, the mechanic said that actually we had fixed it rather better than the ordinary factory fixtures. Clearly we had overdesigned it and used more wire and muscle power than was strictly necessary.

By the late summer we had the big job of installing the huge radar set, 277, with a 4 ft 6 in dish as its aerial. It must have been around September that we actually got the set going and there were many months of observation to be done. My two civilians, the invalided soldier and the elderly one, certainly did not feel they could stay up there in winter, when it would be necessary to walk up and down, carrying whatever supplies we needed. Interestingly enough the Admiralty thought it right to replace those two civilians by four sailors; fortunately there was room for them all. We also, of course, had to try to get all the necessary supplies up while the railway was still open. That meant fuel for the diesels and for the central heating and hot water, and all the food we might require, and the like. It was a major undertaking but in the end we managed quite well. We certainly brought up plenty of fuel, we were always nice and warm, although the fierceness of the climate there is considerable. The aerial was designed for conditions in the Arctic for, of course, it often went on ships that took convoys to the Soviet Union. It had a 28 kilowatt heater installed to prevent ice formation. When conditions were severe on Snowdon, you could not actually tell from looking at the icing of the aerial whether the heater had been on or off. However, while it was on it was easier to knock off the bits of ice with a broom handle.

The work went well and I am proud that the earliest sound and comprehensive measurements of wave clutter are mine. In late January I felt the work had been done, and there was no point in staying on. But how were we to evacuate the place with so much secret equipment? I negotiated with my headquarters and finally arranged that we could leave everything there, except the handbooks of the radar set, which explained everything about it, and the magnetron valves, the powerful transmitter valves, which were one of the secrets of the British radar success during the war. These valves were awkward devices, massive copper blocks with glass protrusions sticking out of them, evacuated of course. Each of them could only be carried safely suspended on springs in a wooden crate. Carrying equipment had to be manufactured by us, in the form of straps so that one could take these crates on one's back. In addition I told the sailors to manufacture little rucksacks for themselves, for their personal gear. And I then worked out that the five of us could just about manage everything in one go. The only risk and worry were the gales, because a winter gale on top of Snowdon is not to be taken lightly. In the gust of a gale it was wholly impossible to stand up. Also, the mountain was covered in thick snow, and to find the route required reasonable visibility. I was in radio telephonic contact with the nearest airfield, and so asked the Meteorological officer there to let me know, during a certain week at the end of January 1945, on which day I could expect the likelihood of gales to be minimal. It was, I think, on the Wednesday, that in our morning talk he said that this was the day. So we began the lengthy task of closing down the place. This meant letting the central heating go out. It meant stopping the diesels which, by that time, were so worn out that I had little confidence of being able to start them again, and, above all, to let the water out of the large static water tank, which had been our main water supply for, of course, in our absence it would freeze and burst, and in the spring, when it melted again, would cause devastation. It was maybe eleven, when all this work was done, and the weather no longer looked too good. But we just had to carry on with the plan. So, after a rapid little snack, we set off. Very soon visibility deteriorated, but fortunately my bump of direction was good enough. The four sailors were all battle hardened but, as I had noticed before, the mountains scared them stiff, particularly one of them, who was the cook, a heavy awkward young man. I therefore gave him the easiest lot to carry, the bag with the handbooks, while the other three sailors and I, in addition to personal gear, carried the magnetron valves. I made the footsteps in front in the snow so that the others could follow. About half to three-quarters of an hour out from the top the gale struck; there was nothing to be done in the moment of a gust but to throw oneself on the ground and let it pass, before one got up again and took further steps along the route that I had chosen, which effectively was the railway track. People who know the area will recognize that on the descent, before very long, it crosses the steep slope of Crib-y-Ddysgl, the mountain next to Snowdon. I must confess that I was a little worried about the risk of avalanches, because it was not very cold and so there was no reason to feel certain that the snow was stable. Therefore I tried to get my people to spread out a little. Through one of the gusts when we all had to throw ourselves down I heard a scream from the back. I turned round and the clumsy cook had thrown himself on the ground so roughly that the straps of his makeshift rucksack had burst, and the bag full of secret documents was sliding down the snow slope out of sight into the mist. There was nothing I could do about this at the time, and I may say that I was mighty glad when I got all my little party otherwise safe and sound with all their equipment down to the bottom. Of course I reported the events to my headquarters, adding my private comment that I was sure that enemy agents would find easier places to pick up handbooks of this set than the snow-covered wild mountainside of Snowdon. They would have none of that. They were absolutely furious. A senior official came to North Wales post haste, mobilized the Commando Training School in the next valley, and two days after the event I found myself leading a group of maybe thirty-five Commandos to the site where the bag had slipped out of sight. I must confess I was more worried than ever. It had not snowed, true enough, so the bag would not be covered. But it had got even warmer, visibility was still poor, and the danger of avalanches clearly had increased. But of course these Commandos from the Mountain Training

School did not know the very concept of being afraid. And so they scampered happily all over that mountainside, no doubt rather skilfully. And, indeed, before very long they located the bag which had become lodged under a rock and was found safe and sound. So we brought this down in triumph, but I do not think I was ever again in good odour at my headquarters. This was the rather dramatic end for my one period when I was actually working on experiments rather than theory.

Coming back to Witley, I threw myself into the normal work there, still very active, though we could all see the end of the war coming.

The other side of my work in 1943-5 was to prove the foundation of my whole future. Fred Hoyle was always full of ideas in astrophysics, full of problems with which neither Tommy nor I were acquainted. When he stayed with us at Dunsfold, which was during the week in many periods, we talked until late at night about these questions. He and Lyttleton had started on a new line in astrophysics, abundantly justified since. Up to that time the general feeling had been that space between the stars was normally empty and that the system of stars, or galaxy, was basically rather old and unchanging. They started from the opposite end, that the system was still in a rapid state of evolution, an evolution caused by the large amount of gas and dust the presence of which they postulated but which equipment at that time was unable to detect. In particular, a central question was the possible accretion of such material by stars. Hoyle and Lyttleton had made soundly based suggestions about how this might happen and what the rate of accretion might be. But they had not gone into detail and these ideas were not widely accepted. The problem interested me mightily. Partly it led back to the kind of hydrodynamics I had worked on in 1941-2, but even more so to my efforts in the Admiralty when I had studied how the magnetron worked. It is curious that the theory of how this vital device acted came much later than the experiment, and indeed its development and engineering use. In the compartmentalized world of wartime I worked out the theory of the magnetron, with interpenetrating streams of electrons, but found I had been anticipated, if very briefly, by Büneman, who worked with Hartree in Manchester. This magnetron work gave me vital clues for the study of the accretion of material by stars, a problem I solved in the spring and early summer of 1943, and then wrote up as my fellowship thesis for Trinity.

For those not acquainted with the Cambridge system it is worth saying that the best thing, in my view, that Cambridge does for its bright young people is the system of junior research fellowships whereby such people, often well before the Ph.D., are elected into fellowships on account of the brilliance rather than completeness of their work. These fellowships are enough to live on, though not luxuriously, yet they convey great status. I had been aiming for this before, but realized very well that my submission in 1942 lacked the necessary bite. I can still remember the moment when

I got the news of my having been elected a Fellow of Trinity, with all this meant, at the very beginning of October 1943.

It was Fred's wife, Barbara, who told me of the telegram, when I arrived at their house in Funtington, a little before Fred, on an occasion when we had business down there. At that instant I appreciated that scientifically, I had arrived.

Perhaps it was even more important in those days than it is now. For, in those days, you saw from the title pages of books where the authors described themselves as "sometime Fellow of Trinity College" as something exceeding all the other distinctions they might have acquired later. In my day it was considered *infra dig* to read for a Ph.D. while one was already a Junior Research Fellow, and so I abandoned my registration for this. Of course I came to Cambridge for the Fellowship Dinner at the beginning of the October term 1943. The College, as always, was not only kind, but very pleased that their much earlier hunch about my potential had proved right.

For me it meant that I knew what I was going to do when I came back after the war. I was not to be a struggling research student working for a Ph.D., but an established researcher as a Fellow of Trinity. Much later, when it became my job to see whether people were young enough for Junior Research Fellowships (the usual age limit is 30, but there are invariably people who ask for an extension), I was always rather hardhearted. After all, I had got this Fellowship before I was 24, with the period spent in full-time research being a mere seven months and in fact the work I had done in those months had very little to do with what I had got my Fellowship on. But, of course, my astrophysical work did not end with this success. Fred and I worked further on my thesis subject, extended it and published a paper a little later, in the *Monthly Notices of the Royal Astronomical Society*, where so many of my papers were later published. This was my first substantial paper. And we worked on other problems. Tommy, though not a mathematician, was always contributing greatly with his wonderful physical intuition, while I was the one who did most of the more manipulative work. There was an especially memorable occasion when Fred and I were each working out values of particular terms in a complicated equation in some problem or other, and there was stony silence in the room while his and my pencils wrote furiously. Then I could not remember which side of my equation my term went into. And so the next words Tommy heard was me asking Fred "Now do I multiply or divide by ten to the power twentyfour?" Tommy couldn't contain himself laughing over a situation where I was apparently uncertain of a quantity to the tune of a factor of 10^{48}. And so we did a good deal of work, not as momentous in its consequences for me as the work on accretion, yet very significant in forming my skill at research, in forming my attitudes, in forming my understanding of physics and astrophysics.

With the end of the war in Europe in May 1945, and even pessimists expecting Japan to be defeated within a year or two, there was not much point in my work at ASE. For new developments took at least that time to become useful for the Fleet. Yet release from my war work was likely to be slow. But Newnham College, Cambridge, helped me decidedly by asking me to lecture to their students during the Long Vac Term of July and August 1945. Teaching had a high priority and so I returned to Cambridge in the summer of 1945, six weeks before the war in the Far East ended. As soon as I arrived I was told by my old friend, Besicovitch, that in addition to my Fellowship I had also been appointed an Assistant Lecturer by the Faculty of Mathematics, so that I had an immediate second job and, with it, responsibilities in teaching which I knew I was going to enjoy.

CHAPTER 8

Thoughts on the
Second World War

THE SECOND World War had such an effect on me and on all who lived
through it, on all that has happened since, that I feel it is desirable to
make some personal comments about it. Moreover, it is a great worry to
me that to the generations that have grown up since the Second World
War, its character is to some extent obscure, to them its outcome seemed
to be obvious and fore-ordained, something that we should all have
foreseen. In fact it was not like that at all. I believe the risks of failure
are grossly underestimated by those who have not lived through those
years. Indeed, there was every reason for the opposite outcome to have
been feared and indeed expected. There was first, the unbroken run of
successes of the Hitler regime. In every sphere, political and military,
Hitler was totally successful from 1933 to the early autumn of 1942. In
this period of nine and a half years there was not one piece of business to
which he put his hand where he did not succeed except for the Battle of
Britain in 1940 and his failure to take Moscow in 1941. First, in con-
travention of the peace treaty, he introduced conscription, without any
opposition, and rearmed Germany. Next his troops reoccupied the Rhineland, the demilitarized character of which had been the one safeguard
the Western Allies had to protect the countries of eastern Europe from
German expansion. Again he managed this without meeting any resistance. On he went through the Anschluss of Austria, to the Munich
crisis, to the occupation of the rump of Czechoslovakia in March 1939. In
everything he had tried to do he had been basically successful. During
the war he defeated Poland more quickly than anybody thought likely:
out of the blue he managed to occupy Denmark and Norway, including
the far North of Norway, in a successful manoeuvre of the greatest
boldness. He knocked out France in 1940 in next to no time. The campaign
in the spring of 1941 against Greece and Yugoslavia was without stop or
comma. And, finally, the occupation of Crete was a serious battle, but
again one that he won.

Therefore the aura of invincibility was tremendous. It is true that in
the Battle of Britain he had not managed to destroy the Royal Air Force,

its defensive or offensive capabilities. Nor, very fortunately, had Hitler defeated the Royal Navy's ability to deny his armies the ability to cross the Channel. Had they been able to do so, their eventual success in Britain, after bitter fighting, seemed only too likely. The fear of an invasion of these islands was entirely justified in 1940-1. Also, in the winter of 40-41 the bombing that he dealt out to the United Kingdom was, by the standards of the day, very severe. In the Western Desert the great victory under Wavell of General O'Connor over the Italians was rapidly reversed when Rommel and the German Africa Corps came into action. Yet I may add a footnote here:— it does not seem to me to be sufficiently stressed in many accounts that the victory of British armies in the Western Desert against the Italians in December 1940 was vitally important and far more significant than it is given credit for. In an obvious sense it led to Rommel starting further from Suez than he would otherwise have done, but in a less obvious sense the effect on the United States was of outstanding significance. That winter of 1940-1 was when British money in the United States ran out. Without the Lend Lease Act of early 1941 all the imports would have stopped, which involved essential supplies of food and arms, with the United States by far the most important supplier. The horror of Hitler in the United States was considerable, but so was isolationist sentiment. Many, Joseph Kennedy in particular, regarded Britain as a lost cause, and wanted America to strengthen herself to discourage Hitler from trying to cross the Atlantic, but saw no point in throwing good money after bad, by supporting a Britain which they regarded as a totally defeated country. The brilliant victories in the Desert of December 1940 came just at the right time to change this image. There was fight in the old dog yet. I doubt myself whether the Lease Lend Bill could have passed Congress at all, or certainly not at the required speed, without that important victory. Again here Churchill's greatness shines through. When fear of an invasion of England must have been ever present, to send troops and tanks and artillery around Africa, to be able to defend Egypt and, to order such a brilliant offensive required courage of the highest order. Only if the victory in the Desert is seen in this context can its vital importance be appreciated. But this was victory over Italians, not Germans. As soon as the small German Africa Corps arrived, the situation became much less favourable. Of course it was not just the arrival of the Africa Corps that changed the position. It was the move of many troops from North Africa to Greece that led to the great retreat in North Africa. Churchill has been severely criticized for this move to Greece which, indeed, from a military point of view made no sense at all. However successful the Greeks had been against the Italians on the Albanian border, against the German attack from Bulgaria and through Yugoslavia, they would be helpless. The small numbers of British troops available could be decisive in the

African context, but they could not have a significant effect in Greece. So this expedition was foredoomed to failure. Yet I think politically, it was impossible to avoid at least some attempt at helping the Greeks. After all, Greece was an ally, and in her hour of greatest need, not to move a finger to help a small country that had not sided with Germany, as most others did, was politically impossible.

It was perhaps different with Crete and there was, after all, a real chance of holding that island. Fortunately, in my view, the battle for Crete was another defeat, for this led to a disengagement which was essential for British forces. To maintain an outpost like Crete against constant hostile air attacks would have been an intolerable strain on the Mediterranean Fleet and on the Royal Air Force.

For all that year, from June 1940, and the capitulation of France, to June 1941 and the German attack on Russia, the United Kingdom and the dominions stood alone against Hitler, with his established aura of invincibility. Courage and unity and superb moral leadership were vital. Winston Churchill supplied this leadership, as no one else would have done. My admiration of his unique capacity has been unwavering over the years. Thus when, decades later, I was asked to preside over the national memorial to him, Churchill College, my appreciation of this honour was very deep. Though the internal deeds of the Hitler regime were not liked, nobody could fathom the depths of horror to which they would sink. Indeed, in those days of 1940/41, and much of the next few years, while few doubted the appalling nature of the Hitler regime nobody could think or imagine just how awful its actions were going to be. The first question is, therefore, what would have happened in the world if the United Kingdom had surrendered, or at least been neutralized, during that long year from June 1940 to June 1941. In no such historical question can one give an answer with perfect certainty. But one or two matters seem clear. Any hope, in Churchill's words, of the New World coming in to restore balance in the Old one, would have been foredoomed to failure. Against the ruthless and effective German submarine campaign the Atlantic was kept open with the utmost difficulty, using great forces, naval and air, from both sides. The idea that with the whole eastern shore of the Atlantic under German control the United States, however resolute, could have invaded Europe is sheer nonsense. The surrender of Britain would have made Hitler's western flank utterly secure. Moreover, there would not have been the southern flank, for almost all that was anti-Hitler in the Mediterranean then was British. His campaign against Russia in the second half of 1941 was very successful, but failed to destroy the Soviet Union as a military machine. It must again be a matter of debate how near German forces came to destroying the Soviet Union and its powers of resistance. We know that getting to Moscow in 1941 was touch and go. We know that had he, without his preoccupations in Africa and the Balkans, been able to start the campaign six weeks earlier, he almost certainly would have reached Moscow and beyond. Indeed, when in early 1942 Molotov came to visit London, with the best will in the world Churchill could not express complete confidence that the Soviet Union could survive 1942. At the very least it is arguable that a British surrender in 1940 would have led to Hitler's dominion over all of Europe, including all of European Russia. He would have been protected by an unbridgeable Atlantic from the West, his friends in Japan could prevent an attack on him through Asia, and so he would have dominated all of Europe and much of Asia, for the Middle East and Asia were defenceless without Britain, as was all of Africa. The dream of the song of the young Germans, "Today Germany belongs to us, tomorrow the whole world", was not far from fulfilment. The attempt to secure continental Europe's dominion of all the old world was not foredoomed to failure by any means. Only British resistance and resolution, perhaps greater than he expected, prevented Hitler's war leading to success for him. However wicked his starting of the war, it was neither an insane nor an irrational act.

Churchill, so rightly, described 1942 as the "hinge of fate". America was at last able to throw her power into the scales and succeeded in saving Australia and New Zealand from Japanese by the remarkable Battle of the Coral Sea, and the very bloody campaign in New Guinea and the Solomon Islands. Two or three comments about the Japanese theatre of war are in place. Again the unbroken success of the Japanese in the first few months of their campaign was staggering and mesmerizing. Perhaps to the younger people the idea of a fundamentally not very strong country like Japan being so successful and so feared comes as a surprise. But when Pearl Harbour was followed by the sinking of the two major British naval units in the region, when area after area was overrun, the continued American resistance in the Bataan Peninsula was the cause of the greatest admiration by all but seen as of moral rather than military significance. One feared greatly for Australia and New Zealand, who had helped so enormously in the Middle East and in the United Kingdom earlier in the war. There is a modern tendency to say that they were betrayed by Britain and America, who decided that Germany must be dealt with first. It is said by some Australians that this showed lack of gratitude, lack of concern, and indeed was treasonable. I would wholly disagree with all this. Germany was and, as I will indicate, continued to be a mortal danger to the world as a whole. Germany alone could contemplate separating America from Europe: alone could develop new and fiercer weapons: could gradually bring into play some of the conquered peoples of the Continent on her side. Japan really had too low a base to do this. Thus for the sake of the world as a whole the Pacific war had to be put into a lower gear. This decision threatened a terrible tragedy for two of the most civilized countries in the world. It was avoided by brilliant American campaigns and successes. In spite of giving priority to the German war, they managed to defeat the Japanese in the Coral Sea, at Midway, in New Guinea and the Solomons. These successes in no way detract from my view that priorities were put correctly.

For many months indeed 1942 was a bad year. It started with the unbroken Japanese advances and the fall of Singapore on the 15th of February, which followed close on the heels of the demoralizingly successful German dash through the Channel, with their ships, *Scharnhorst*, *Gneisenau* and *Prince Eugen*.

Here, incidentally, I feel that the tactics pursued by Churchill and British forces were unsound. Given the importance to the German ships of their fighter cover, given that at that stage the RAF had already shown great strength in bombing German airfields in northern France, it would have been better to switch all the attack to those airfields, and thereby prevent the air cover over the ships that caused such heavy losses to attacking forces, and meant that their success was so limited.

When the Germans resumed the offensive in Russia in spring 1942 the one bright spot was that they no longer had the strength to attack on the whole front. However the way they succeeded in the south was really frightening. It was still uncertain whether they would or would not, by these means, succeed in knocking out the Soviet Union. A major question arises here - why did Russia manage to resist so fiercely? After all, it is now known that the Stalin regime was ghastly, so why did people fight for it with that fierce determination that gave them victory? I think the answer is that they knew, as the West did not, how terrible it was to be occupied by Hitler's troops. The wilful, indeed playful, slaughter of unarmed people, the total racial contempt of the occupying forces for what they regarded as a sub-human population are almost unimaginable to us. This experience gave them a fierceness, a readiness for self-sacrifice, a courage and determination that nobody who had not experienced Nazi domination could imitate. It is no criticism of the courage and devotion of the forces of the Western Allies to say that they did not have the same spur of needing to succeed, needing to liberate their country, that was the dominant force in the incredible Russian revival and reversal of fortune. Perhaps the only time in the West that one had seen a similarly incredible performance was at the time of the Battle of Britain when, again, every participant knew that the existence of the country itself was at stake.

Jumping forward a little, what has been called, particularly in America, the betrayal of Yalta, was in my view not a betrayal at all. The dominance of the Soviet position towards the end of the war was due to the incredible recovery and successes of their forces, which the Western Allies, not having experienced Nazi ferocity in their own countries, found it hard to match. Hence the slowness of the Italian campaign, compared with the Russian advances, hence the position of the forces at the end of the war. And diplomacy could do very little to change in peace what had happened in war.

The retreat in Africa in 1942 was, of course, bitter, as Churchill has so much stressed. The fact that Rommel did not get to Suez was an unexpected and most fortunate surprise and was due to Auchinleck's victory in the July battle of Alamein. In October and November of 1942 the great reversal occurred, with the Battle of El Alamein, with the landings in North Africa, just before the reversal at Stalingrad. But it should not be thought that the remaining two and a half years of the war were all plain sailing: that after the North African and Stalingrad successes victory was clear. There were enormous obstacles left, which might have become much greater still. From where I sat in Naval Radar Research, the miracle seemed that the Germans were so late in introducing the snorkel for their submarines. There was no new technology in this; it was all made of bits which already existed. If their submarines had not had to surface at night to recharge their batteries, the submarine war, which was very nasty indeed, would have been very much fiercer still. And even in 1943 and 1944 it might have become possible for them to break the Atlantic lifeline, not only preventing the deployment of American troops in Eur-ope, but starving the United Kingdom into submission as well. So the risks of the Second World War coming out differently from the way it did were not only real, but they persisted for a long time. Without this understanding, neither this autobiography, nor most of the world we live in, can be understood.

A point that I have not seen made by others concerns the origin of the doctrine of unconditional surrender. In the summer of 1940 and persisting for quite a while in Britain's lone struggle there was a strong feeling of moral superiority. Not only was it felt that France collapsed through treason, through lack of certainty about what they were fighting for, but there was also just disgust with all those who made compromises, like Petain and Laval. Ours was a pure cause and we would not sully our hands with compromises, as was difficult to avoid for those who were unhappily occupied by the Nazis. This purity of attitude could not be carried forward from defeat into victory. When, in the landings in North Africa, the American and British forces negotiated with the detested Darlan, suddenly all our purity was lost. This was a major worry for many of us. We did firmly believe that Churchill would never surrender, but was it possible that the American and British governments might eventually negotiate with Nazi supporters as disagreeable as Darlan had been? The shock to the Allied system was very great. And I believe myself that a few months later Roosevelt and Churchill declared the policy of unconditional surrender because they realized the need to regain the necessary feelings of moral superiority that had been lost over the Darlan episode.

Another little story. I had long prided myself on my understanding of geography, of railways and, to some extent, of military affairs. When the Germans, in their retreat in 1943, lost the vital railway junction in Zhitomir, I felt sure that they could not allow this to happen. If they were really so weak that they could not reverse this loss, then the end of the war was near. I made a bet with one of my colleagues that either the Germans would retake Zhitomir within two days or the war would end that year. Since they retook Zhitomir the next day, I won my bet and got confirmed in my belief that my understanding of military matters was quite good.

Nobody who lived through the war can ever forget it. The world we know was shaped by it. Victory over Hitler was essential, at any price. Rarely, if ever, was there a regime so utterly evil with the will and the power to spread its vile doctrine of German racial superiority over a large part of the globe. There have been wicked governments before and since but none that posed such a danger to the world. For example, we know of the horrors Pol Pot committed in Cambodia but there was never any danger that he would spread his tentacles further.

The world that emerged from victory over Hitler was by no means all good, but the one that would have arisen had he won would have been utterly horrible. Nor was high drama absent from the endeavour to destroy Hitler's regime. It had been wholly and uninterruptedly success-ful from its accession to power in early 1933 to the Battle of Britain in summer 1940, and generally successful for a further two and a half years until Alamein and Stalingrad. The permanent fear even during the long retreat thereafter was of new devilish weapons appearing that could revive its fortunes. The V1 and V2 were bad enough, but could not prevent Germany's collapse. But there was a real worry that nuclear weapons might appear from this land of evil, a worry that did not disappear until the regime was utterly destroyed. Then and only then could one heave a sigh of relief.

CHAPTER 9
Life in Cambridge after the war

I WAS ONE of the very first of the younger Fellows to return, partly because of the fortunate circumstance that Newnham had asked me to give a course of lectures during the long vacation in July 1945. My friends drifted back slowly on their release from the Forces, or on their release from war work. Of course, many of the older people were in Cambridge and I had a very agreeable time with them, as a young Fellow, enjoying the port in the Trinity Combination Room. And the older members of the community thoroughly enjoyed having young members amongst them again. Of them I must particularly mention the Master of the College, George Macaulay Trevelyan, who had become Master following the death of J. J. Thompson in 1940. Trevelyan was a historian, not only of great distinction, but one who could write and talk about history in the most lively and enthralling manner. All my old interest in history was re-awakened and I owe very much to him, particularly for the periods on which he has written so brilliantly, the formation of Italy, and England under Queen Anne. I have enjoyed history ever since and have had sympathy for the subject, though of course I certainly have not had the time, even if I had had the inclination, to become a professional.

Another historian at the time, who was very agreeable and kind to me, was Kitson-Clark. Other friends from before the war were around, notably Besicovitch, Dean and others. Outstanding among the great people in my field, or rather the adjacent field of pure mathematics, there were in Trinity G. H. Hardy and J. E. Littlewood. I never got very close to Hardy who was beginning to be unwell, and died shortly afterwards. But I became good friends with Littlewoood, essentially because we both enjoyed drinks after dinner. The situation in Trinity then was in the Combination Room that two wines circulated, port and claret. The iron rule was that no two bottles could ever be mixed. Therefore, if you finished a bottle at your place you got a full glass of the new bottle, the so-called buzz. Littlewood was always very late for dinner in Hall, because he did not want to be present at Grace, another point that engaged my sympathy, and so he was usually also rather late coming upstairs to the Combination Room. He often stood for some time in the doorway before he decided where to sit down, which, quite often, happened to be next to me. At first

I was intensely proud that this great mathematician should so much enjoy my conversation that he came to sit by my side; it was only later that I fathomed what he was doing when he was standing in the doorway. He was trying to calculate who would have port, who would have claret, where the buzz would occur and accordingly seated himself in order to get the extra glass. It had nothing to do with my personal charm. Yet we continued to get on very well.

Some years later he and Mary Cartwright did a fabulously involved and interesting and unexpectedly successful piece of work in the finest tradition of analysis, on a differential equation which was of some importance in electrical engineering. They then decided to write it up so that it could be understood even by the least mathematically educated, so that the electrical engineers would benefit from it. But they felt they should try it out locally on the least educated person, to see whether this was really true and for this purpose they chose me. I might mention here that when I tried to understand this wonderfully involved analysis, the point that I least understood is how they could have imagined at the beginning that there would be a way through this enormous tangle, and then Littlewood told me a story I have always treasured. It tells of two mice who fell into adjacent buckets of cream. They both struggled hard on and on to avoid drowning. Then one gave up, said "It's no use" and died. The other kept on struggling. Suddenly the cream turned to butter, the mouse jumped out and lived happily ever after.

Another very interesting member of the society of Trinity in those days was Bertrand Russell. I am a great admirer of his views on religion, of his mastery of philosophy, but much less so of views he had in mathematical physics, which was my subject. He was always interesting always provocative, not necessarily very likeable with his high-pitched voice and his argumentative bearing. One of his great ideas of the day was that the United States should make preventive war on the Soviet Union to destroy its capability to build atomic bombs. It is an interesting sidelight on Russell's impeccable logic, that then, as now, what he saw as the ultimate disaster, like most of us, is war between nuclearly armed powers. He then thought that one way of avoiding this was to make sure that there was only one nuclear power in the world. However repulsive it would have been to attack a country that had suffered so much during the Second World War, that did not seem to him to weigh in the balance. The moment there was more than one nuclear power in the world he switched completely and became a total opponent of nuclear weapons. This attitude for which he is so well remembered was for him a perfectly logical sequel to his previous attitude, but a change that few others could have followed.

Gradually my own friends of my year returned. John Pinkerton became a research student, and later carved out a career for himself in computing. John Kendrew, who had had a splendid career in Operational Research, with Bernal and Blackett, soon went into the brand new subject of Molecular Biology with Max Perutz, founding a new science that was to bring Nobel Prizes to both of them, and to quite a number of other members of their team. Jack Gallagher came out of the Army as a Private. We all had been sure that he would come out either as a Private or as a General, depending on whether his brilliance or his unorthodoxy would impress seniors more. Apparently he had twice been in Officer Cadet Training Unit, only to return to the Army as a Private. He now worked in Imperial History, particularly in relation to India, a field in which he was to make a great name for himself, and he continued to be the best and most interesting of companions. Stephen Ainley came back briefly, again a splendid friend and companion.

But what was professionally most important for me was the return first of Fred Hoyle and later of Tommy Gold. Fred, some four years my senior, married with two children, was already oppressed by the price of houses in Cambridge and got himself one before long, well to the south of Cambridge near Quendon. All of us worked to get the house into a good state and went out there on many occasions. Fred, on the other hand, spent his days in Cambridge, making my rooms his headquarters, so that I saw a great deal of him, and there I was also joined by Tommy Gold who, at that time, became very much interested in the theory of hearing which he developed and this brought him also into contact with Jerry Pumphrey, with whom we had worked so closely during the war. His brilliant theory of hearing is, thirty or forty years later, beginning to be understood, but it was quite extraordinary how the specialist fraternity turned away from what even they, much later, had to admit was the right solution.

I had also got back into playing Bridge, of which I had done a fair amount in internment, and I had very agreeable Bridge companions, consisting of Maurice Pryce, back doing Physics in Cambridge, though before long he moved and later was for many years at the University of British Columbia. Eric Wild, who somewhat later went to Manchester, and Sam Devons, who went to the United States and taught at Columbia University made up the foursome. Though we were all on the physics side, it was Bridge that kept us together, and we became quite good at it. I enjoyed the game considerably. Though a mathematician is supposed to be good at chess, I was never any great shakes, though Stephen Ainley and I played each other on occasions, but never in any way even approaching expertise.

I thoroughly enjoyed my teaching and my lecture courses. I was especially keen on a course I gave on functions of Mathematical Physics which had previously been given by pure mathematicians. I changed it considerably to fit my applied outlook, and it became an extremely popular course. And I think, quite generally, my courses went well with the students. Supervision I mostly enjoyed. I liked having to deal with examples. I greatly enjoyed going through these with good and medium level students, but when they were really incompetent I found it a bit trying. Perhaps because I was a little younger, perhaps because I was rather good at reading the questions carefully, I became of great help to Fred too, who found getting into the swim of these exam questions rather harder than I did, and he dubbed me the "Supervisor's Supervisor".

Before long I was also involved with University exams which I modestly enjoyed. I thought setting questions of the right level, and convincing one's fellow examiners that they were right in their standard, was good fun and agreeable. Marking exam scripts would have been great pleasure had one dealt with modest numbers of, say, thirty or forty. But the mathematics classes at that time were rising towards the hundred mark, and that became a bit of a bore, though with only a few questions per examiner this never became a great strain, as it did later in my career, when I was in London.

Perhaps it is worth mentioning here another examination endeavour, in which I became involved before long and which was to prove quite a source of irritation and a bit of an eye opener. Trinity had quite a number of entrance scholarships and exhibitions for young people at school wanting to join us. And the selection was made by means of an examination. Of course, with Trinity's standing, a good few of the candidates were in mathematics. It was very often the younger Fellows, of whom I was one, who were pressed into service there. It was a most awkward examination in that no printed syllabus existed. One had to be guided entirely by what happened in previous years. To everybody's irritation this gave a premium to good drilling at school and it was a severe lesson I learned then. The Entrance Scholarships and Exhibitions in Mathematics were won by boys from schools good at teaching the mental gymnastics of these questions. However, only some of them became excellent mathematicians and only a few of these youngsters, who had performed so well in their exam that they swept the board, ever became a Fellow at Trinity. The drilling had clearly enabled them to pass the exam well, exam successes which greatly aided the standing of their schools with parents, but had driven out all love of the subject for some of these young people. I should mention that one of them became in fact a very good professional mathematician, John Shepherdson, whom I have already mentioned before. But even he did not get a Trinity Fellowship, though he did become a Professor at Bristol. But it then became quite clear to me that judging schools by their exam successes does not tell one a lot about their educational standards. Indeed, as I knew from my friend John Shepherdson, the general education of boys trained for the entrance scholarship exam was often poor.

But above all there was my research. Stimulated for a while by Harold Jeffreys, no longer my research supervisor, but a mathematician of great standing, by Lyttleton, by Gold, by Hoyle, there were any number of problems I worked at, and the papers came out at a very considerable rate.

My appointments were, of course, limited in time. My Junior Research Fellowship at Trinity lasted for four years from my return to Cambridge, that is until the summer of 1949. My assistant lectureship came to an end a year earlier, but I never had any doubt that at that stage of my career, the right place for me to be was Cambridge. And though various temptations were put in my way fairly early, to go to Manchester, to go to London, as a Reader and the like, I firmly resisted. Indeed in 1948 I was appointed a University Lecturer which is a permanent position and my base was secure. There was more of a gap in my Fellowship of Trinity, though a past Fellow enjoyed considerable rights. By the time my fellowship had expired I had got married and lived out of College. Yet the status was agreeable. However, not until a vacancy arose among the staff fellowships that are given for teaching, could I be re-appointed, which happened in 1952.

My researches covered a very wide area. Fluid motion, waves on the surface of water, geophysics, the solar corona, rigid body mechanics, electromagnetism, were all topics I worked in and published on. But perhaps there are some research efforts I should talk about a little more fully. I became known in wide circles as a result of the publication of the Steady State theory of the expanding universe in 1948, with Tommy Gold, and the virtually simultaneous publication of a similar paper by Fred Hoyle. It may be, therefore, worthwhile talking about this at some length. But it should always be seen in context. Though this was an effort in cosmology, it was only one of many things I did, and I always detest being referred to as a cosmologist. Though I got fame from that subject it was far from being the only field in which I did research. The cosmological scene at the time was indeed puzzling. The great name was Edwin Hubble who, in the twenties had established first that the nebulosities in the sky were distant galaxies, much like our own, but definitely outside our galaxy, contrary to what had been assumed by many earlier. And secondly, that there was in operation the velocity distance relation, which was the basis of the notion of the expanding universe. In this kind of motion the velocity of recession of a galaxy as inferred from the red shift is proportional to its distance. The red shift is directly observable. By a well known relation in physics that is turned into velocity. The distance of an object is far harder to infer and it is invariably a long and involved chain of argument that gets one there from the raw observations. But what was clear from the observations was relative distances; the red shift of the objects was so correlated to their apparent faintness, that when the faintness was interpreted as due to distance and red shift turned into velocity, it was indeed a law of proportionality. But the absolute value of the distances was much more difficult to estimate. The constant ratio of distance divided by velocity (which is a time) came out at 1800 million years in Hubble's estimate. Now compared with what was already then known about the age of the oldest rocks on Earth, and some observations about the ages of meteorites, this was an embarrassingly short time. But the authority of Hubble, rightly based on his brilliant work in establishing the extra-galactic nature of the nebulosities and the velocity distance relation, made it difficult to doubt his estimate of the characteristic time of the universe. This caused the oppressive time scale problem that lasted throughout the 30s and 40s. It led to great activity in unorthodox approaches to cosmology, such as those due to Milne and to Dirac, as well as to efforts in the more orthodox manner associated with Eddington, Lemaitre and Gamow. Hoyle, Gold, and I had discussed these issues amongst others on many occasions. But it was in, I think, the second half of 1947 that Tommy Gold came up with the thought of continual creation and the steady state. Fred and I thought this was a crazy idea which we could shoot down before dinner. Dinner was rather late that night! Very soon we found out that not only could we not shoot down that idea, but that it dealt with many other problems. We had all, I perhaps more than the others in the field, been impressed by the thermodynamic characteristics of the universe. We were struck by the oddity of what seemed a rather old system having such enormous temperature differences as between the hot surfaces of the stars and the very cold space between them. So we gradually saw that this idea not only solved the time scale problem, and the problem of the origin of the universe, but also was a major contributor to understanding its thermodynamic condition. Moreover, Tommy and I, perhaps a little more interested in the philosophy than Fred, were very much struck by the logical desirability of this theory. By Popper's criteria, being the most vulnerable theory it was therefore the best, the one that gave the most definite statements that could be tested. The steady-state theory cannot take refuge in explaining any inconsistency by referring to changes in, for example, the nature of galaxies between past times and the present. According to the theory whatever went on anywhere at any time must go on here and now. Thus our own theory provided a much more beautiful and predictive picture than anybody had had before.

We then ran into a problem. The theory as such was just an idea, and ideas cannot be published by themselves. Fred solved this problem by rushing into a field theory. He, being a Field Theorist, found this convenient; Tommy and I, not being so wedded to that side of things, felt that it was premature with a theory that tried to account solely for the large scale features of the Universe. A field theory, we argued, was only needed to show how the theory dealt with variations between different areas. Fred decided to publish independently, but his publication, for various reasons was delayed. Tommy and I felt a bit stuck, until one evening I discovered that Hubble's counts of numbers against brightness fitted our theory very well. This was the peg we hung our publication on. But it is really rather ironic, because I am the one who later not only was most critical of Hubble's number counts, but of anybody's number counts. Yet in the interests of historical veracity this must be mentioned. Our paper appeared almost simultaneously with Fred's. All three of us, particularly perhaps Fred as the eldest, became fairly well known as a result of this theory. And today many people connect my name still mostly with this theory.

The theory caused a bit of a scientific sensation and much argument. The established opinion favoured an evolving universe, starting from a state of very high density, the "Big Bang". However, there was no observational evidence against the steady state model in those days, though the discussion could be quite heated (see p. 90). In the 1950s the theory stimulated the analysis, by Hoyle and colleagues, of the origin of the heavy elements which is universally accepted. In the 1960s evidence began to accumulate (the microwave radiation and the abundance of helium) which made most astronomers accept the Big Bang model though this too has some problems. (However, since the mid-fifties I have not worked in cosmology.)

My interest in cosmology had a little earlier led the Council of the Royal Astronomical Society to ask me to write as one of their Council Reports a review of cosmology. This was published at a not very different time from our paper but, of course, was a broad summary of what everybody did, not a propaganda exercise for our particular theory. Not much later Cambridge University Press asked me to write a book on cosmology. It is certainly the longest book I have ever written. It was eventually published in 1952, and I prepared a second edition for publication in 1960. But it is widely regarded as having been very influential in setting out the problems and their possible solutions in cosmology. It was a big effort for a young man, and I must confess that to this day I am rather proud of this book. However much science has moved on, it is still considered a classic of its kind.

Another paper of the time that might be worth mentioning. Following on the work I had done with first on my own and then with Fred on the accretion of matter by stars, when the star was in motion relative to the cloud of matter, one afternoon I decided to look at the problem when the star was at rest relative to the cloud. It turned out to be a very simple analysis that, within a few days, I had completed. I regarded it as so simple, not much more than an examination exercise, that I did not want to publish it. My colleague Lyttleton persuaded me to publish. Quite a few years later this whole field became very active, and this paper now is a citation classic, having been quoted in innumerable papers by other scientists. That just shows how poor one's predictive abilities are.

After this digression to science I should return to my personal life. Having been separated from my parents and sister through all the years of the war I was, of course, keen to pay them a visit in New York. This was not too easy in the still very disturbed conditions of the time. Moreover, I was still a stateless person. But I did manage to get on a boat to the United States in the summer of 1946, and thoroughly enjoyed my visit. It was marvellous for the family to get together again after all those horrific years. New York in August is of course not normally the best place to go to, but that summer was a very easy summer there when it really was not uncomfortably hot. Late that autumn of 1946 I became a naturalized British subject, and the issue of a British passport was going to ease travel very much. Indeed, Tommy and I and many other young people in our circle were thinking that the spring of 1947 might be a good occasion to go for a skiing holiday on the Continent. Although we were naturalized (Tommy achieved the great step at virtually the same time), the issue of a passport required the Naturalization Document to be registered by the Home Office. The Home Office, true to its usual character, said this would take about three months. Fortunately we had in mind combining the skiing trip with a visit to a conference that was vaguely connected with Sir Edward Appleton, and the mere mention of his name was enough to cower the Home Office into registering our certificates speedily and so we got our passports quickly.

This had very major consequences for me. Fred Hoyle was always a very elusive person when he was not in my room. And in trying to find him in St. John's College, or in the Art School where lectures were given, I made contact with a young research student of his, Christine Stockman. Before long we decided we both wanted to go on this skiing trip and the skiing trip was not all that far advanced before we felt we might want to get married. Skiing, of course, is well known to be a dangerous activity in this respect and indeed our party resulted not only in our marriage, but in that of Tommy Gold and of Nick and Margaret Kemmer. Kemmer was also at Trinity, a rather older and very well established theoretical physicist, who later became Professor at Edinburgh. The skiing was wonderful. We went to Arosa where we rented a sizeable chalet for a good size party. Amongst other people there was Freeman Dyson, a brilliant mathematical physicist, a little younger than I, who not much later moved to the United States, first to Cornell and then to Princeton, where he still is. He is a very amusing character, an excellent dinner companion at Trinity. Two incidents about him I particularly remember. At Arosa where we wanted to be a little careful with food, he was the only man I ever saw frying up the previous day's porridge, while if we sat together at high table (he became a Fellow at Trinity even younger than I) we had so much to talk about that on one occasion he stopped me and said "Not another word now, we're still on the soup, and other people are being served the sweet." It was a splendid skiing holiday, eight years, or rather more, since I had last had my previous one, twenty years before I was to have my next one. Our marriage plans matured. The suggestion of our marriage, as between a Jew and a non-Jew, caused some raised eyebrows amongst our respective families. Their worries rapidly disappeared once there was personal acquaintance. As for Christine and myself, our firmly non-religious views, both in respect of ourselves and of the upbringing of our future family, was a perfect protection against the divisive influence of religions!

On coming back to Cambridge, Christine and I developed our plans. I was of course very keen then that my parents and sister should meet her, and so another trip to the United States became desirable. Finding room on a ship now became markedly easier, though it was still rather expensive for our not-very-good incomes. Christine of course was worried, having heard how dreadfully hot the whole East coast of the United States was during the summer, and I put her fears to rest by my experience of the previous summer which, unknown to me, had been so exceptionally mild. And so we sailed off on an American ship, largely taking troops on leave, the *Marine Tiger*, for a not over comfortable trip. That summer the United States East coast was desperately hot and, of course, air conditioning was not yet common in private homes or in private cars. It really was pretty dreadful. We fled to the White Mountains of New Hampshire with my sister, where we went to stay in Pinkham Notch. In those days one still travelled by train, indeed by sleeper, which was comfortable and air conditioned. But I still remember the shock of getting out of the sleeper into the heat of late morning at Intervale in New Hampshire. High up at Pinkham Notch it began to be easier, and then when we really went walking in those lovely mountains, with their beautiful woods, the temperature was very agreeable, especially after a major thunderstorm had changed the atmosphere a little. Nine years earlier, in 1938, that area had been hit by the great New England hurricane, and the shock of coming round the side of a mountain to see an exposed face and to find woods still devastated in the aftermath of this great gale is clear in my mind.

New York society was, at least in those days, still very tribal. My parents and sister lived in the Jewish and, to some extent more specifically German Jewish circles, a segregation that to us was totally strange and difficult to fathom. My parents also came up to New Hampshire but to a bigger resort, Bethlehem, where we stayed for a few days, before returning to New York. With both of us interested in astronomy, we managed to meet Chandrasekhar on Cape Cod. We then had a good passage back in the *Queen Elizabeth*. My parents never were really happy in New York. Of course I had strongly advised them not to settle in the biggest city in the country, but that was not an advice that they felt able to follow. Though with the relative value of pound and dollar my father's income seemed to us fabulously high, my parents, and in particular my mother, felt constantly oppressed by what she called their poverty. Of course the different social circumstances, where domestic labour was incredibly expensive and where one used modern machinery entirely strange to them to make life easy, passed them by, and they, in particular my mother, never understood the significance of the Victorian dictum "If you look after the pennies the pounds will look after themselves." So it was always a rather oppressive atmosphere there which, together with our dislike of big cities, made going to New York a sacrifice. Christine and my parents got on splendidly and they were immensely pleased to gain such a charming and impressive new daughter-in-law, though their persistent shortage of money for major expenditure meant that neither of them could come to our wedding at the beginning of November 1947. My mother, always very proud of her son, like any good Jewish mother, was thrilled to the core to find that we had got together as equals. On the other hand she had not yet quite taken in the measure of success in my career. I remember when we left, she said, "Become a great man" and Christine immediately replied, appalled, "But he is a great man." Meanwhile Christine also had had her major success in being elected to the Isaac Newton Studentship, a most prestigious studentship for her work in Astrophysics. This greatly eased our otherwise very tight financial situation. My father, through an artist friend, had a connection with Einstein. This enabled me to pay him a visit in his Princeton house during our stay in New York. For me as a very young scientist our meeting, lasting towards an hour, was a most memorable occasion. He discussed with me various problems relating to particle motions in an atmosphere of total equality, as one scientist talks to another. Thanks to my father's connection I also own a splendid line portrait of Albert Einstein which always hangs in my study. I proudly lent it to the Royal Society in 1979 for its Einstein Centenary celebrations.

I should here mention another figure in Trinity who looked after us with the greatest kindness, the Senior Bursar, Tressilian Nicholas. Not only was he kind to me as a young Fellow, but today I owe particularly much to him, for he advised me, and I followed his advice, to start a pension scheme in my non-pensionable employment as Junior Research Fellow and University Assistant Lecturer. This early start pays off handsomely now in terms of the pension I enjoy. Equally he was the Administrator of the Isaac Newton Studentship, so we were two of his charges. Not long ago, at the time of writing this, he passed his 100th birthday, still as keen on music as ever, still a very loyal member of the College. (He died recently, aged 101.)

I should also say a little about Christine's family. Her father was a man of remarkable ability (especially in mathematics) who from very humble beginnings rose in the Civil Service to the rank of Deputy Secretary. After his retirement he was very active for the International Labour Office. He and his wife resided on ILO's behalf first in Barbados and later in Geneva before finally returning to England. Very active in old age, he died rather suddenly in 1982 when he was 88. My mother-in-law came from Scotland, only survivor of a family of three, her brother having lost his life in the First World War and her sister having died young in the 1919 influenza epidemic. Her cousin was a delightful lady who lived with them. Also a civil servant, she died in the early sixties, whereas Christine's mother lived to 1989. Christine has a younger brother, Michael. He and his wife are veterinary surgeons (he was at one time President of the British Veterinary Association). They are good friends.

We got married on my 28th birthday, the 1st of November 1947, in the Registry Office at the Shire Hall, Cambridge. It had been very clear to us from the beginning that neither of us wanted a wedding that was religious in any way. Trinity had already begun the policy of buying up property adjacent to it in Trinity Street, and making it accessible from the rear to accommodate its students, since accommodation in town was getting more and more difficult. Of course whenever in Cambridge I had lived in College since the end of my first year as an undergraduate, when I became as an exhibitioner, a member of the Foundation. For my last two years as a bachelor Fellow, I had a lovely set on the East side of Great Court. It so happened that different levels made it difficult to make the corner house at the angle between Trinity Street and Trinity Lane accessible from the rear. So this was converted into two flats above the shop on the ground floor, and we took one of those, again with help and advice from Tressilian Nicholas. It was but a stone's throw from my previous Fellow's rooms, so I always claimed that I got married very mildly. We stayed there for nearly a year and a half to the spring of 1949, and Fred Hoyle and Tommy Gold, who lived a little to the south of Cambridge, both found our flat as good a Cambridge central headquarters as my rooms in College had been, and discussions went on all day, every day. But Christine had also interested me in her work on the structure of the stars, and we did a good deal of work on this together, and there are quite a number of joint papers by us. And so my work continued to be extremely fruitful.

In the summer of 1948 the International Astronomical Union met in Zurich, in Switzerland. We were both members of the Union and got some help with travel, which enabled us to start with a holiday, beginning in Geneva and continuing in the very beautiful and then still very small resort of Saas Fee, where I had been in the last summer just before the war broke out. From there we went to Zurich, and encouraged by my rather senior friend Bill McCrea, talked about the steady state theory informally to a number of senior astronomers a few weeks before it appeared in print. It was the first of many meetings of the Union I attended.

My mother came to visit us also that summer of 1948, in the second half of which Christine became pregnant. With a child coming, we wanted a house and I had become established as a University Lecturer that summer of 1948. We found a charming small house at the end of a complicated little street on the north side of Cambridge, maybe a mile and a half from the College. I remember my more mathematical friends, on their first visits, saying to us that early on they decided that in that street our house was either the last one to exist, or the first one not to exist.

Since I had been a boy and met my cousin Thea, ten years younger than I, at the age of 2 1/2 and found her utterly charming, I had been very keen to have a family; inability to have a family was for me the worst nightmare of the future. So when our first one was born I was absolutely in seventh heaven and then gained a piece of education that I have treasured all my life. While Christine was still in hospital I had to have my hair cut and, as happens in Cambridge, in the chair next to mine was a senior revered member of my profession, Professor R. O. Redman. He heard me gushing about the joys of being a father, and then interrupted me and said, "Now look, we know all about this, we've got four. And there are two things I can tell you about children. First, they grow up just as they want to, and there's nothing you can do about it, and secondly, from the time the first is born, at least until the youngest is 16 (which is the stage they had reached), the house is sticky from top to bottom." I was much impressed, but just how true those words were only became clear to me as the years rolled by.

The move to our house had one very serious consequence. Fred Hoyle, who had been to see us almost every day, now found that he had been robbed of his most convenient headquarters in the middle of Cambridge, and it was several years before he came to see us again. Scientifically I was now much more on my own, though close links with Tommy Gold continued. But this in no way diminished my scientific output; I was now quite self-propelled, though then as always I enjoyed scientific discussions. In that lovely summer of 1949 when Alison, our first one, was very young, we had a visit from my sister from America, but took no holiday, and in the summer of 1950 we joined the Golds at a wonderful, but very primitive spot, Westmill, Welcombe, in North Devon, at the very point where Devon joins Cornwall, a wild and beautiful area. But it was an exceedingly wet summer, in a primitive house, not the ideal place with two children, ours of 15 months and theirs a few weeks older. Early that autumn I realized I had been serving quite a while lecturing in Cam-

PLATE 1 Visit to the Meteorological Office, Summer 1973. H. B., Administrative Officer
Meteorological Office, Lord Carrington (then Defence Secretary).

PLATE 2 Photograph of Einstein drawing referred to in text.

PLATE 3 Chiefs of Staff Committee 1972. Head of table: Admiral of the Fleet Sir Peter (now Lord) Hill Norton CDS. On his left: Admiral Sir Michael Pollock CNS, then Air Chief Marshal Sir Denis Spotswood, General Sir Michael (now Lord) Carver, CGS, Sir James Dunnett, H.B., Almost opposite CDS (in dark suit): Sir Patrick Nairne.

PLATE 4   Sydney 1965? Professor Harry Messel presents H.B. with lambskin at
the end of H.B.'s lecture series.

PLATE 5 Institute of Mathematics and its Applications, Presidential Dinner 1977, H. B. (President 1974–5), Duke of Edinburgh (1976–7), opposite him Dame Kathleen Ollerenshaw (President elect).

PLATE 6 Meeting at Liege, Belgium, 1952.

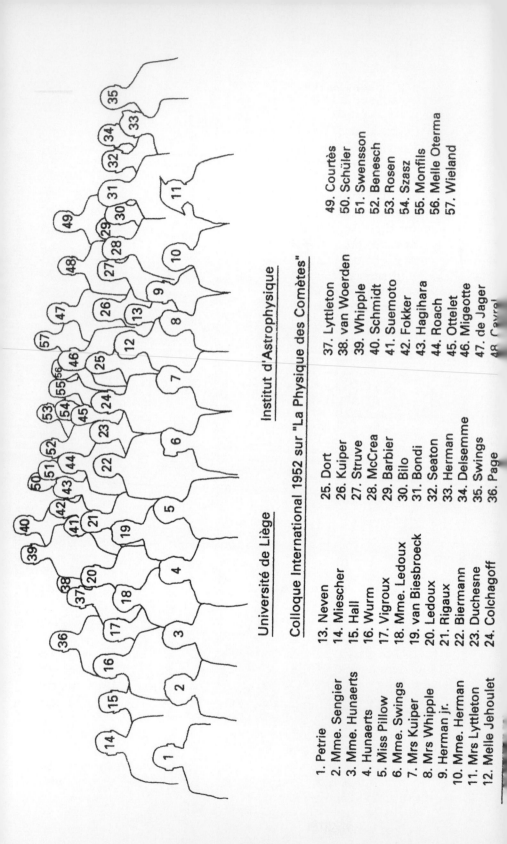

Université de Liège

Institut d'Astrophysique

Colloque International 1952 sur "La Physique des Comètes"

1. Petrie
2. Mme. Sengier
3. Mme. Hunaerts
4. Hunaerts
5. Miss Pillow
6. Mme. Swings
7. Mrs Kuiper
8. Mrs Whipple
9. Herman jr.
10. Mme. Herman
11. Mrs Lyttleton
12. Melle Jehoulet

13. Neven
14. Miescher
15. Hall
16. Wurm
17. Vigroux
18. Mme. Ledoux
19. van Biesbroeck
20. Ledoux
21. Rigaux
22. Biermann
23. Duchesne
24. Colchagoff

25. Dort
26. Kuiper
27. Struve
28. McCrea
29. Barbier
30. Bilo
31. Bondi
32. Seaton
33. Herman
34. Delsemme
35. Swings
36. Page

37. Lyttleton
38. van Woerden
39. Whipple
40. Schmidt
41. Suemoto
42. Fokker
43. Hagihara
44. Roach
45. Ottelet
46. Migeotte
47. de Jager
48. Coyral

49. Courtès
50. Schüler
51. Swensson
52. Benesch
53. Rosen
54. Szasz
55. Monfils
56. Melle Oterma
57. Wieland

PLATE 7 With Prince Charles at the naming of the Royal Research Vessel Charles Darwin, early 1984, Appledore, North Devon.

PLATE 8   With General Goodpaster, U. S. Army, Supreme Allied Commander
Europe, at SHAPE, 1972.

PLATE 9 At RNEC about 1973. The man on my right is Ian Davies, then Assistant Chief Scientific Adviser.

PLATE 10   Washington 1975 with the Chairman of the U. S. Chiefs of Staff
Admiral Moorer.

PLATE 11 At a U. S. Military Establishment, Yuma, Arizona 1976.

PLATE 12 At British Aerospace with Sir George Jefferson (later Chairman of British Telecom) and Sir George Leitch (Chief Executive, Defence Procurement).

PLATE 13 European Space Research Organization (ESRO) Council 1970. Second on my left is Professor Puppi, Chairman, and third is J.-A. Dinkespiler, the Director of Programmes.

PLATE 14 At the Guildhall of Cambridge 1985

PLATE 15 Touristic moment on the Defence Visit to India, January
1973 with Ian Davies (Assistant Chief Scientific Adviser) and John
Elliot (my private secretary).

bridge, and so could claim a term's Sabbatical leave. We naturally wanted to combine this with seeing my parents and for this, and other reasons, an American university was the obvious place to go to, as one could generally get there into some paying position. I wrote to a number of places and got a number of forms back, but from Cornell I got a personal letter from the great physicist Hans Bethe, not only warmly welcoming the idea of my being at Cornell as a visiting research fellow, but ending by saying, "I gather you have a wife and young family, so would you like to stay with us as our house guest while you look for a place of your own." This, of course, was irresistible. The forms all went into the waste paper basket, where all forms basically should go. And in March 1951 we embarked on the way to New York. It was a dreadful crossing, made worse for Christine, always a bad sailor, by being pregnant again; made particularly difficult because the little child, Alison, now 21 months old, got over her seasickness much more quickly than I did and was of course keen to be entertained. I stayed briefly in New York with my parents while Christine and Alison were there rather longer. I found a very nice apartment in the home of people called Smith, in the lower part of Ithaca, the University being higher up the hillside. It was a most beautiful area and scientifically excellent, and I received much stimulus and pleasure talking to my colleagues there. However, there were already worries. This was the period of the beginning of the McCarthy troubles and there was serious concern about the future of academic freedom in the United States. Though the visit had begun with Alison getting the measles, after that it all went extremely well and we thoroughly enjoyed ourselves in Ithaca, a place very much to our liking (though it rained an awful lot). We were in the curious position of living in America without a motor car, and neither of us, at that stage, could drive. With only the experience of my non-driving parents in New York and the town situation there, we had not become really clear about what a difficult thing it was to live in America without a car. So many of our friends offered to lend us their second car, but we just had to say "No, thank you."

After this fruitful stay we returned to Cambridge and work continued well. One other little anecdote. Being an untidy mathematician requiring a lot of paper, I always used to work on the floor, sitting there cross-legged, a practice I continued well past middle age. But of course, when Alison began to walk about this created havoc. So we put her in a playpen, at which she screamed. The solution was then obvious: I went into the playpen with my papers, and this worked extremely well. Our son Jonathan was born on my birthday in 1951, exactly 32 years younger than I and just four years after we got married. So we have a triple celebration on that day every year.

Work continued apace. I started to have research students and a particular association then began which still continues. Felix Pirani returned to England from a lengthy stay in North America, where he had worked with my old friend and companion from internment days, Alfred Schild. He arrived with excellent recommendations for his mathematics and equally good ones for his baby sitting, so we exploited him quite a bit at that stage. His career and mine have remained academically closely connected.

My work continued well, but whatever I experienced in Cambridge reinforced my dislike of administration. Whether in the College, which was very competently looked after by its Council and its splendid Senior Bursar, or the University, where a short period on the Faculty Board did not give me the feeling that I was any good at this type of work, nor that the administration worked particularly well with the academics. Of course, in a subject where one's whole equipment is pencil and paper one can keep away from administrative work as one cannot in an experimental field. We had been charmed by the United States and we were now both able to drive and enjoyed the freedom a car gave us, particularly with two children. We had a very good holiday in South Devon in the summer of 1952, and then began to scheme for another visit to the United States. In the summer of 1952 the International Astronomical Union had met in Rome. This was after some severe political troubles, the meeting having originally been planned for 1951 in Leningrad, then cancelled at the turn of 1950 in the wake of the disasters of the Korean War. This caused great and, I think, entirely justified resentment on the part of our Soviet colleagues. The choice of Rome in a NATO country did nothing to assuage this. However, the Rome meeting was organized, the Soviet scientists agreed to come, and it promised to be a good meeting. So I wanted to go, but money was a major concern. Therefore I eagerly accepted Lyttleton's invitation to go with him and his wife, and Bill McCrea, by car. Of course, roads were poor in those days, and I got my first leisurely acquaintance with France and Italy on that trip. In Rome I also met many senior people, which was highly useful in our next endeavour to visit the States. Being now so wedded to a car we were greatly intrigued by the idea of driving across that country, along a southern route, in particular visiting the astronomical observatories of the southwest, and ending up in California. This time the choice fell on Harvard, and so we went there in September 1953. Again while Christine and the now two children stayed behind with my parents in New York, I tried to find accommodation in the Harvard area, and found a lovely apartment in the beautiful little town of Concord, Massachusetts. A very senior member of the Harvard Observatory, Fred Whipple, looked after us extremely well and so we had a happy three months in the autumn of 1953 at Harvard: to be followed by the long trip to California, long before anybody dreamt of interstate highways and the like.

But meanwhile another event had happened in my career. In the early summer of 1953 I received a request to visit Kings College, London, to apply for the vacant professorship of Applied Mathematics. I had really no reason to go. I was now a Staff Fellow of Trinity and a university lecturer, both tenured positions giving one great comfort. But there were beginning to be stirrings of unease in me, and perhaps more in Christine. Yet when I took the train to London that day I think I viewed it purely as a courtesy visit. The delightful atmosphere I found at Kings, largely created by the Principal, Peter Noble, and the Professor of Pure Mathematics, Jack Semple, made it a very different proposition. Indeed the College made it so clear to me that I was very much wanted that the matter had to be thought out very seriously. When my wife and I discussed it, various points came together. We were both keen to live more rurally with our family, but in those days in the Cambridge countryside professionals were almost unknown. If there were any there, they were almost bound to be academics. In the commuter belt of London there were lots of professional people of many different kinds, and so a much broader society could be expected. I had loved Surrey, its hills, woods and commons since my war time experience at Dunsfold and Witley, and Kings' was readily accessible from that side. Lastly in Cambridge one was only too well aware of people who had spent all their lives there, who were like the trees and the furniture, and who had no experience of the world outside Cambridge.

We felt the time had come to make a move from Cambridge. However, the vacancy was for October 1953, whereas we were already fully advanced with our plans to spend September 1953 to April 1954 in the United States starting. We also thought it would take a little while to sort out our affairs in Cambridge, and so it was agreed that I would take up my new Professorship only in October 1954. It was against this background, knowing that when we returned it would be for a major move, twelve months ahead, that we embarked on our American trip. For the first time we crossed the Atlantic by aeroplane, and found this a far superior experience (though it was still in the days of propeller planes, Constellations, needing two stops, Prestwick and Gander). Our stay in America was the first one when we had a car; we bought a 1949 Ford. This car enabled us to taste life well away from Harvard, where I taught and researched   at the Observatory. In the delightful small town of Concord, Mass. where we lived our eldest, then 4, went to a nursery school. New England in autumn with the magnificent colours of the foliage is most beautiful.

At Harvard I met many senior astronomers of great distinction. Harlow Shapley, recently retired as Director, was still much in evidence. His successor, Donald Menzel was most pleasant, as was our sponsor and host, Fred Whipple and his wife. Bart and Priscilla Bok became good friends. They moved to Australia not much later. There were many excellent and promising young astronomers there who much later made considerable contributions to astronomy. A younger astronomer who became a great friend and who we still see from time to time is David Layzer. We travelled a certain amount in the Autumn term, between Concord and New York, and we also took a trip to Pittsburgh, where Alfred Schild was, and then to Ithaca to meet our old friends of 2 1/2 years earlier. Being overtaken by an early snowfall, we discovered the problems of travel in a snowy country with two small children. And then, in January 1954, we started on the big trip. Money, of course, was not plentiful and so, with a lot of organization, we had managed many visits for modest fees to Universities and Observatories, often with hospitality included. To travel through the old South, and then the Southwest in 1954, out of season, was a remarkable experience. The industrialization of the old South had not really begun, and in many ways it was what one would now call a developing country. The last motorway ended seventeen miles south of Washington, after which we saw no such road, at least outside the towns, until we came to Texas. One of my memories of those days is stopping for a night in a tiny place in Alabama, where there was a tiny motel, in which all four of us stayed. In the morning when I wanted to pay, I had run out of change and had nothing smaller than a twenty-dollar note (of course twenty dollars in those days was worth a good deal). This produced total consternation. "Oh gee", said the motel owner, "I haven't got enough money to give you change for that. Now George at the grocery next door, he had that much and more yesterday, but then he took it to the bank", and it took a real whipround in this little community to give me the change for my twenty dollars. We next visited friends from Ithaca days in Baton Rouge, at Louisiana State University, and then went West, stopping at many observatories along the way to California. The desert country was entirely new to us, but none of us took to it all that much. I remember little Alison saying that really she thought that a daisy was much nicer than a cactus. Yet we also got the chance to see the Grand Canyon, and to experience a very different world. In California we were again in the midst of plenty, both by nature (there were still orange groves everywhere before the flood of housing overtook them), and money, and science. And we managed to visit, if briefly, both Sequoia and Yosemite Parks. Innumerable incidents and impressions of this big trip could be told, but I will confine myself to those associated with Lick Observatory and the wonderful Director there, Douglas Shane, and his wife, who were such splendid hosts to us. We had arranged to go there on a March Sunday afternoon. The day was very dull indeed and we were surprised and bothered by the amount of traffic on the very narrow, winding mountain road leading up to Lick and wondered whether perhaps we had timed our arrival stupidly, when some major event was going on. Not at all. But in the poor weather a little snow had fallen near the top, and all the citizens of the town at the base wanted to go up and see the snow and show it to their children. Fortunately there was a police control before the most difficult part of the road starting near the summit, when only people like us, with business on the top, were let through. Mrs. Shane had amongst other modern kitchen equipment an electric blender and an electric waste grinder (Dispose-All). When she used her blender to make especially good soup from the day before's left-over vegetables, she referred to it as "Cream of Dispose-all". While we were there the yoke of the big new telescope they were building came up this mountain road, and I still remember our two children, their noses glued to the window, watching the manoeuvres, as the huge vehicles came up this very difficult and awkward road with their vast piece of equipment.

CHAPTER 10

King's College London

GETTING HOME in April 1954, there was one more term's teaching in Cambridge, and then the task of selling the house and of finding a home for the future. This we aimed to be well away from London, but with good train connections to a terminus reasonably near King's. Also we wanted to be in an area of my beloved county of Surrey. My colleague Jack Semple lived in Redhill, and he and his wife suggested I might start looking for a suitable spot from their house. Our first idea was to build. We had been so impressed by the prefabricated timber housing we had seen in America that we wanted one like that. Discovering that there was a firm in England making these with imported timber for export to Australia and Canada, we got them to design one for us, for use in Surrey. With a little struggle we managed to get over the planning problems, but the mortgage problem proved insoluble. It was not a question of fire risk; the extra fire insurance was minimal. It was just a question of attitudes. When I asked a building society why they would not give us a mortgage on this, the only reply that I could get out of them was that it would not be a house that would be easy to sell, because our successors would not be able to get a mortgage! The circularity of this argument did not seem to impress them at all. While all this was going on, of course we had to live somewhere. Renting property then was, perhaps, not quite as difficult as now, but yet I was surprised to get very few replies to an advertisement which I placed in *The Times* for renting. One of the replies turned out to be acceptable but this was a not very wonderful house on Reigate Heath. The position of the house, though, was simply superb. I had never known that one could live on a heath and not on a road, and fell in love with this kind of situation. Yet the house was difficult to heat. It was clearly not ideal for us. And so, when eventually the last of our plans for a prefabricated house fell through, we had to look for something else. By that time, we were well and truly settled. As soon as we knew we were going to live on Reigate Heath we started to look for a school for our eldest. She was 5, and started school in Cambridge, but after one month of this we moved to stay with the Golds at Herstmonceux (Tommy was then Chief Assistant at the Royal Greenwich Observatory) and after one month of the school there moved to Reigate Heath. Thus a permanent school for her was essential.

The nearest school to us was in the village of Buckland, but this was a Church of England school, as was the school at the closest end of Reigate. We of course much preferred a County school. A little further was Holmesdale School. When I went there to talk to the Headmaster, Cliff Price, we took an instinctive liking to each other, which started a long association. For the next seventeen years until July 1971, Cliff Price, whose period as headmaster came to an end then also, had Bondis on the roll, usually two, for a brief period three, occasionally only one. It was an excellent school and we enjoyed sending our children there. Moreover, in the, to us, new locality it was through the school that we made friends with other people bringing their children to school. So we found good friends and companions for the many years we lived in Surrey. We found the area most engaging and liked the people there. And so we stayed in Surrey in fact for thirty years. Of course, school is all very well, but the first year a child goes to school it brings home a lot of infections readily picked up. Our house, as I was saying, those first two winters was somewhat cold and damp, and the children were ill a great deal. We had the most punctilious service from our doctor, and he or his partner came whenever we called. Indeed, they were so much used to coming to this slightly out-of-the-way spot, that on one occasion one of them came without having been called.

The children grew out of this and we managed. While we were in this house, in the summer of 1955, our third child, Elizabeth, was born. That summer, too, a house on the other side of Reigate Heath came up for sale, and though originally we thought we could not possibly afford it, in the end we managed to make a deal. The history of the house was quite interesting. Towards the end of the nineteenth century when the daughter of the squire in the big house on the hill got married, he quickly put up a little shack for the couple that would do them before they had much of a family, so he only put eleven bedrooms in. By the mid-1950s this of course had become an unusable place, and a builder bought it, tore down the entire centre part, and offered the two remainders for sale. The south-west-facing one, having the reception rooms of the old place, he sold readily enough. The northeast one hung fire, and so our rather modest bid succeeded. It was at the stage when it was rather in pieces. Therefore I was able to give my interest in domestic engineering free reign. I decided on the heating system, making sure that the northeast aspect need not mean that the house was cold. I designed the kitchen, the general layout, and how we could all fit into this house. I had always enjoyed using my understanding of physics to work out good heating systems, and already had had one put into our house in Cambridge, which I designed. The plumber who put it in, didn't at first like coming under such tight control by an outsider, but eventually reconciled himself to the idea with the remark that two heads were better than one, even if both were sheep. In our Reigate Heath house I had much more scope; one innovation I put in, which I am surprised has not caught on widely, was to use the central heating pump to circulate against gravity and have a non-return valve. Therefore, when the thermostat does not call for the pump to operate there can be no circulation at all. It was a system one never had to turn off in summer and never had to go into a fuss about starting up again. There was no stray heat, and so it was both convenient and efficient. Similarly, I was very proud of my kitchen design which was in the shape of a U, in what used to be the workshop end of the house, with a cooker at the bottom of the U, and a gate at the other end, separating the working part of the kitchen from the meals corner. A child could therefore never get close to the hot end, while yet being in constant conversation with its mother who was working in the area. We had come anxious moments during the rebuilding operations, when the builder threatened to go bankrupt more than once. But, thanks to an excellent lawyer and a most helpful surveyor, in the end it all came good and gave us a splendid family home in the most lovely surroundings.

So many people have the idea of Surrey as a continuous suburb that they find it hard to imagine how we lived. We could not see another building out of the front of the house, that was the open country of the heath, nor at the back of the house, where we looked out over the paddocks of a riding stable to farm land, though admittedly there were two houses by the side. Main drainage never reached our house and I looked after the septic tank for our twenty-eight years there. For the first twenty years of our time there, the electricity supply was very irregular and unreliable. By opening up loft space I managed to give each of our eventually five children a room of their own, while I still had my study which also served as a guest room. After fifteen years there we made some changes, including a bigger kitchen, and a games room outside, but this house continued to be our home.

Throughout our years there, the education of our five children proceeded at the local state school. After the primary stage, at Holmesdale, the three girls were selected for the Girls County (Grammar) School, the two boys for Reigate Grammar School. When the youngest entered the sixth form, the County School changed to a Sixth Form College (which she much enjoyed) while the boys' school became independent, not to our pleasure, just after our younger son left. This interest in state education led to my wife being a founder member (and for a number of years chairing) the local Association for the Advancement of State Education (CASE), of which I was an active member. After initially being very doubtful about the move to comprehensive schooling, we later switched to support it entirely. I, for one, could not see how a divided system could supply the required number of young people with high educational achievement without wasting scarce highly qualified teachers (e.g. of mathematics) on teaching very small numbers in the non-grammar schools. Of course we never were under the illusion that going comprehensive is always easy and successful. But many issues other than this concerned CASE, and it was a very active and useful organization.

Meanwhile professionally things were going extremely well. Felix Pirani joined me at Kings College before so very long. Clive Kilmister was already there, and so we had the beginnings of a group in the subject of the theory of gravitation (general relativity). In 1955 I was invited to a small conference (I basically disliked large conferences and enjoyed small ones) in Bern, to celebrate fifty years of relativity. The great physicist, Pauli, was in the chair, and it was a most excellent meeting. But perhaps it was particularly memorable for me because of discussions we had at that meeting on gravitational waves. The mathematical and physical complexity of Einstein's theory of gravitation is so great that there was still confusion, and a variety of opinions, about whether the theory predicted the existence of gravitational waves or not. After one of those discussions, Marcus Fierz, Professor at the ETH, the federal technical university, took me aside and said, "the problem of gravitational waves is ready for solution, and you are the person to solve it." This remark governed a sizeable slice of my scientific work for many years, and led to my 1962 Paper on Gravitational Waves in a fifteen paper series on Gravitational Waves. The 1962 Paper I regard as the best scientific work I have ever done, which is later in life than mathematicians supposedly peak.

But other important things happened to me too. When I accepted the Chair at Kings, one of the essential draws was that my wonderful colleague, Jack Semple, would be Head of the Department, and I would have *no* administrative work at all. And then, rather rapidly, all this changed. First there came the question of computers. It became clear that the University of London had to do something in this field for its members. At one of the Constituents' Colleges somebody very confident was building hardware. But in my view the time was ripe for the building of computers not to be an academic enterprise, but something for the commercial world. It was expected that this group that was working on hardware would automatically become responsible for the provision of computing for the University. But I had my doubts about this. Therefore I rang a friend of mine in the computer world, and when I mentioned the name of the person who led that group, he said "Ah yes, he has half-built more computers than anybody else." This was decisive for me and I weighed in and said that we had to go commercial, and buy equipment from an industrial firm. The administration of London University, with which I had not really had any contact before, turned out to be absolutely delighted to get a lead. I got all the support one could hope for, and London was set on its way in computers, albeit not without troubles such as delays in delivery, etc. Suddenly, in a major administrative decision, I found myself not only involved but successful and applauded at being successful. Everybody had expected the arguments to last for years before anything was done. Due to my intervention they were over almost before they had started.

My verbal ability to sum up speedily and easily also came to the fore. In those days the various Colleges of the University took what was essentially the same examination. Many matters of common concern in each subject were dealt with by a Board of Studies, meeting perhaps twice a term. The meetings usually took two and a half to three hours, and were atrociously boring. When it became my turn to chair this body, the meetings reduced to forty minutes or less, took the decisions required each time and, as far as I could discover, nobody felt at all resentful about the firm leadership given.

Finally, there is what I like to call the story of my complete demoralization. One day, at my home in Reigate, there arrived a fat envelope from the Secretary of State for Air, asking me very courteously to sit on the Meteorological Research Committee. Never having had any connection with meteorology I thought this was absurd, but took the precaution to telephone somebody I knew from wartime, who was serving in the Meteorological Office. "Ah yes," he said, "actually I was present at that discussion, where it was thought it would be good to have an outsider on the committee." So I said yes, and took this to be a licence to make a fool of myself, which I proceeded to do. I am never very good at keeping my mouth shut, but on those occasions when I am positively invited to contribute I can be very prolific. After a short time my interventions settled down in about the following proportions: in about 60% of the cases, the rest of the committee made it clear to me that what I had said was nonsense, about 30%, or rather more, where they were all convinced it was nonsense, but did not succeed in convincing me, and the modest remainder where my remarks were actually of some use. After two or three years of this I had a very bad patch, where the 60% grew to 90%, and the useful contributions became very few. So when one morning as I was rushing out a fat letter arrived from the Secretary of State for Air, I could not even be bothered to open it, I thought I knew what it would say — that he was grateful for my services, but unhappily they had embraced the principle of rotation and so they would now have to do without me. But, when on my return I opened the letter it asked me to become the Chairman of this committee. My original academic inclination to think that I could contribute only in something of which I had real mastery was finally shattered. I could say and do things in areas of which I knew little, and apparently it was also appreciated. Since then I have accepted many calls on my time and energy even well outside any area in which I had expertise.

An interesting academic task also came my way. The University of London was active in getting going universities in the Colonial territories, and I was asked particularly to look after mathematics and physics in the University of the West Indies, in Jamaica, where I went twice, and I also paid brief visits to Ibadan in Nigeria and to Legon in Ghana.

In summer 1958 the International Astronomical Union met in Moscow making up for the abortive plans for 1951. In 1958 the Soviet Union was still pretty well inaccessible for the ordinary traveller, and so both Christine and I, as members of the Union, were very keen to attend the meeting. Christine's mother was willing to come to our house for the two and a half weeks we were to be away and look after the then four children (aged 1–9) provided she was given some help by another person. We asked King's College whether there might be a student prepared to do this as a vacation job, and were sent a delightful young woman, Jean Walker. So all was prepared for the great trip. Our friend David Layzer from Harvard stayed briefly with us on his way to Moscow. In five minutes it became clear that Jean and he had fallen for each other. Not so much later they married, and we now have this couple (with four children) as splendid friends to visit whenever we go to the Boston area or when they come to England. That they found each other in our house has made us very close ever since.

We flew to Copenhagen, took a boat to Helsinki, and the train to Leningrad. The beauty of the city and the richness of the Hermitage impressed us deeply, as did the then evident poverty of the population and the reminders of the terrible siege it endured 1941–4. We took the train to Moscow where our meeting took place, lavishly hosted by the USSR Academy of Sciences. The opulence of the reception in the Kremlin was striking, and contrasted distinctly with conditions in the town.

Though the size of the Assembly (it was then about 1200 strong) was greater than I like, yet it was an excellent opportunity to hear some of the latest results in astronomical research, to meet our many friends in the community, and to make new ones. The meeting took place less than a year after Sputnik, and the justified pride of our Soviet hosts in their achievement and its scientific results shone through everything.

Two little stories of the occasion stick in my mind. Our being bad sailors had made the voyage through the Baltic disagreeable, so for the return journey we changed to flying from Helsinki to Copenhagen. When we had arranged our air tickets with the young Swede who ran the SAS office in Moscow we asked him how long he had been stationed there. "One year two months and eleven days" was his answer, which spoke volumes about the joys of living in Moscow.

The Assembly extended well over a week. On the Sunday in its middle we were taken for a tour of the huge industrial and agricultural exhibition north of the city. After being guided through one pavilion, Christine,

David Layzer and I had enough of being guided and decided to return to the city. However, we noticed that there was a pavilion of the wine-making industry and chose to visit it. We found the main entrance closed and entered through the side entrance where we found most elevating statistics of wine growing in the different republics. Very soon we were leaving, but when we said yes to the question "Astronom?" we were guided through corridors to a large room where we were sat at a little table and invited to taste a series of a dozen wines and several brandies, clearing our palate with caviar. Though, having arrived a little late, we had to work hard, I looked round, recognized a few guests and realized we had strayed into an occasion for the heads of standard institutions of the world whose modest sized meeting was simultaneous with our huge astronomical one. When the time came for speeches it fell to Sir Gordon Sutherland, Director of the National Physics Laboratory, to reply for the guests. After thanking our hosts and saying what a splendid event this was the went on to say that as soon as he realized what the occasion was like he had looked at his watch, for he knew Bondi was in Moscow and wondered how long he would take to join the party. Thus was my reputation in those days!

At King's College my teaching of course essentially involved lecturing, but to quite small classes, so that it was always a very personal relation. The groups were only twenty to twenty-five, and I had my own style of lecturing without notes. I was put nicely in place by one of our students one day, when he said "It's wonderful to be lectured to by somebody like you. It would be dreadful if everybody followed your method of lecturing." I think I understood what he meant: that my lectures were good at inspiring people, but not necessarily of direct benefit in passing exams. Indeed my lecturing style had already, in Cambridge, led to a situation where in a Third Year course, when I espied somebody who I thought was not interested in this particular subject and asked him why he came, he responded "I came to take a lesson in mob oratory." This got through in other directions too. I was asked to lecture in various schools, other universities and the like, on an occasional basis. The U.S. Air Force got to hear of my research group in Kings College and, having decided for whatever reason to support academic research in gravitation, was generous and allowed me to have a secretary, to do some travel on their budget and, above all, to invite people from elsewhere. This especially allowed us to open links with Poland.

Leopold Infeld, one of Einstein's close collaborators was working there in much the same field as we were, which was indeed not a popular field in the world in those days. Warsaw was perhaps the only big group outside the United States, where at Syracuse and at Princeton there was a good deal going on. It had been my deliberate choice in a small College to start on a subject that was not popular all over the world as we could hardly expect to compete with the big battalions. We then developed very close relations with Infeld's group, with their people coming to stay with us, and our people to them. I myself visited Poland several times in the course of this collaboration. 1960 saw us on another trip to the United States, back to Cornell. Things were much easier by then. The McCarthyism that had darkened our stay in 1951 and especially in 1953 and 1954, had been consigned to the dustbin of history. But I must say I still have, from those days, a soft spot for the very few American universities that stood up against McCarthy, namely Harvard, Cornell and Stanford. Again it was a most enjoyable stay. By now our family was complete, the last two having been born in our house, David in 1957, Deborah in 1959. On both occasions the midwife and the doctor were a little slow in coming, and I took a minor part in the delivery of these two babies.

Meanwhile, things with my parents had not been going so well. My father, whose heart had been giving some cause for concern since 1940, died early in 1959. He had practised as a doctor to the end, well beyond his 80th birthday, which is really not bad going. My mother had been very badly hit by an event in 1956 where her beloved brother, Seppi, an amateur archaeologist, was caught in an incident on the Israel–Jordan border and lost his leg. She had always had a horror of maimings and was very greatly bothered by this. Her health was gradually failing and on our visit in spring 1960 Christine and I and our two eldest children saw her for the last time. She died early in August. However, the link with New York was not broken. My sister Gab still lives there. Very active as a qualified doctor, she spent much of her career in public health, working first for the City and later for the State of New York. (At the time of writing she is in her mid-seventies, but still in full-time work). She married in May 1951 a rather older man, Dolph Lobel, who died some twenty years later. Since we were in Cornell at the time, we could come to her wedding where our eldest daughter, not quite two and very fond of anchovies, pinched them off the top of all the canapés. They had one child, Bernard, a little younger than our middle one, who works for the City of New York. We had quite a few holidays together mainly in the Alps, but also in the United States. But not only do I see them if business takes me to that side of the States, but Gab in particular always spends a little time with us when she takes her own holidays in Europe.

In the spring of 1959 I was elected to the Royal Society, unhappily too late for my father, who would have appreciated very much hearing of this honour. Also at that time I went on a lecture tour of Australia for the British Council. Not only did I make many new friends there, notably Harry Messel in Sydney, the extremely active Head of the School of Physics, but also renewed old friendships there. Bart and Priscilla Bok, whom we had met in Harvard in 1953, were now in Australia, he being the Commonwealth Astronomer on Mt. Stromlo outside Canberra, and this marvellous couple gave me a home from home on Mt. Stromlo. Lord Carrington was the British High Commissioner then. He gave a lunch for me. This was our first meeting and later I worked closely with him.

Other friends in Australia were Eric and Stewart Barnes. He is a pure mathematician of distinction. We had seen a lot of them as they had lived near in Cambridge in the early 50s, and he was also a member of Trinity. They now lived in the delightful city of Adelaide where he was Professor of Mathematics. A former student of mine and her engineering husband lived in Perth, a young astronomer I knew was in Brisbane, and so it really was a most enjoyable trip, though it involved my being away from home for over six weeks. I have been back to Australia many times since but always rather briefly.

Another major task that came my way concerned King's College itself, where I was given the task of producing an acceptable plan for its rebuilding. When I joined King's College London in 1954, it was still in its buildings of the 1830s and 1840s, though a number of houses in the adjacent Surrey Street had been taken over and were in use by the College. There had long been discussions with the University Grants Committee about whether we should move out of this cramped neighbourhood of the Strand, or stay there as a College in the middle of London. Eventually the UGC pledged itself to make us more comfortable on the existing site. This effectively meant taking over all the buildings between Somerset House and Surrey Street, particularly along the Strand, and turning these little old houses into suitable purpose-built accommodation for us. How was this to be done when we already occupied much of this area and our work as a College could not be interrupted? I talked to everybody, analysed the problem, and then made a plan on the simple principle that the first building should be the one where the loss of area due to demolition of existing properties least affected us by being modest and involved the more mobile departments, while the gain in accommodation from the new buildings there should be very much larger than the initial loss. Clearly it was therefore necessary to stage the rebuilding, not only for financial reasons. I was also keenly aware that the availability of public finance for such big undertakings could not be guaranteed in the long term. So equally I was determined that the staging should be such that if at any point we had to stop, the gain already achieved would be very considerable. With a little logic and diplomatic talking it was not really very difficult to gain agreement on a plan, though to turn that into a design and into contracts of acceptable size took somewhat longer. It was a satisfying experience for me to get this whole thing going in my beloved King's College and as the years rolled by my plan was gradually realized. An interesting part of it all concerned a bit of politics. Very early on in my endeavours we were told that London Transport, as it then was, were promoting a private Bill in Parliament for the building of what they called then the Fleet Line, the western and northwestern part of which eventually became the Jubilee Line. On the original concept it was to run along the south side of the Strand to Fleet Street and beyond. This private bill gave them enormous rights over precisely the area we were trying to redevelop, and would have made our rebuilding extremely difficult if not impossible. We approached London Transport for a change in this and were unsuccessful. The only thing left for us then was to appeal to the Private Bills Committee of the House of Commons, and it was my task to represent the College there. The Committee consisted of seven M.P.s whose names frankly I had never heard before (or since). They listened with the utmost attention to my presentation of why the Bill, as it was drafted, was unacceptable, to London Transport's rejoinder, to my comments on that rejoinder, to London Transport's final submission, in a procedure extending over two or three sittings. They then went into a meeting of their own and came up with an excellent compromise, which both London Transport and we could readily accept. I was deeply impressed. Here were our politicians, much vilified by the media, performing a first rate piece of public service. There was nothing in it for them. It was such a dull business that even the local constituency newspaper would not have devoted a line to it, yet it was their attention, their understanding, their compromise that avoided an awkward situation. I have never since been willing to join any politician-bashing argument.

A big job I also had on my plate in those years was that from 1956 to 1964 I was Secretary of the Royal Astronomical Society. My predecessors had had to work extremely hard, as I suspect my successors did. I was extremely fortunate and made use of my opportunity. Just at the time I took over we had selected a new executive secretary as the head of the Society's staff, Mr. E. C. Rubidge. He and I worked together very well, as I did with my colleague, Michael Ovendon from Glasgow. Mr. Rubidge was the full-time person in charge of the other staff, and he could devote all day, every day, to running the Society and its very substantial journals. A very able and effective person, he who knew precisely when he had to consult Michael Ovenden or myself. From him I learned the vital principle of management that one should never keep a dog and bark oneself. I scrupulously refrained from barking and he enjoyed the job that way. It was very well done, and I could confine my attention to a few critical issues, of which there was no scarcity.

Another area in which I got very much involved was the question of a Southern telescope. I have already mentioned how on my first visit to Australia Bart and Priscilla Bok had been such wonderful hosts for me. Bart, in particular, had impressed me with the richness of the Southern skies, which was why he, an optical astronomer, had gone south. Equally I had myself experienced the superb radio astronomy that was flourishing in Australia. British astronomers were quite convinced that Britain needed an observatory in the southern hemisphere. This view was shared by several of our European neighbours, and there was a great deal of discussion as to whether we should join our friends on the Continent in their plan of putting up a telescope in South Africa. I was deeply opposed to this. First, South Africa was already seen to be an extremely uncomfortable place to be in politically. Indeed, one's expectations of its future stability were perhaps unjustifiably poor. But, above all, at that time when I started being involved, South Africa was still a member of the Commonwealth and at meetings with our European friends one could not be explicit about one's doubts about South Africa. But much more than that, I was convinced that it was right to put a great optical telescope, not only where the politics gave one no concern, but where there were great indigenous astronomical efforts. It is so often in the collaboration between optical and radio astronomers that the best advances have been made, so I felt very much that Australia was the right place for us to go, in spite of the inconveniently long and expensive journey. Also Australia had a splendid university system, so that this was not a piece of scientific colonialism, where one placed a telescope in an area where there were not many local astronomers themselves. It took me quite some time to make headway and the matter was, of course, complicated by our relations with our European neighbours. There was a strong undercurrent of wanting to have closer relations with them, something with which I totally agreed in principle. But it seemed equally to me that the size of the British Astronomical community, especially when joined to the Australian one, were such that it would occupy a telescope completely. I persisted and before so very long the Anglo–Australian plan won. Things were going quite nicely when at a late stage the Australians hesitated, preferring an arrangement with the Americans which, of course, was quite understandable, because of the much greater experience of astronomy and of telescope building in America. The American side, however, was much less certain about whether they could muster the necessary finance, and there was a lengthy delay, very aggravating because good opportunities like this, if not exploited, simply disappear. I intervened in America with people I knew there, asking above all that they should make up their minds. They immediately understood this and came to the decision not to support a United States–Australian telescope, and so the Anglo–Australian one was agreed. This, like other difficult astronomical questions involved me much in the bodies that dealt with them, first the relevant section of DISR, then that of the Science Research Council.

In Australia I had made friends with Harry Messel, the immensely active Head of the School of Physics in Sydney, who in particular organized science schools for school leavers, largely Australian, but with contingents also from the United States, the United Kingdom and Japan.

These were made glamorous occasions, in part by asking lecturers from far away. I was one of their favourite lecturers and, if I rightly remember, I went there in 1962, 1963, 1965, 1967, 1970, 1973, 1979 and 1983, always having a very enjoyable time, because nothing is better than to talk to, and then be cross examined (there was always time for this) by these very bright young people. These Science Scholars are selected to be the brightest and most interested in their age class (17–18), and they certainly are fun. Though my time was always very limited, I managed to make an interesting if brief stop on most of these trips. I have been to the Hawaiian Islands, to Mexico, to Cuzco in Peru, to Singapore, to Bangkok and to Kathmandu. All this time I was an active teacher at King's College, lecturing to undergraduates, having research students, and running what became a very famous research school in gravitational physics. I then first became aware that I was really unusually tough and had more energy than most. Colleagues in similar positions to mine did nothing outside their University work, yet had problems with their health from overwork, like heart attacks, whereas I was always in splendid form. I was also very active in the central bodies of the University of London, being for several years a member of the small inner body of the Court, on which ordinary professors rarely sit. My opinions were valued, I took part in discussions on difficult problems and my ability to formulate questions clearly was much appreciated.

Two particular non-University activities of that period I want to mention. The first one was the London flood barrier. The great East Coast floods of 1953 which London escaped, more by good luck than by good management, had alerted everybody to the great danger to which London was exposed if a high tide coincided with a massing of waters in the southern part of the North Sea through the effect of North winds. A Cabinet Committee was set up under the Lord President of the Council, major engineering consultants were involved, yet after ten years all that had been produced were reports from engineering consultants that were strong on contradicting each other, but not very good on anything positive. The question of whether a barrier was necessary with its inevitably high costs, and what the justification was, were still in a very muddy state. It was Sir Solly Zuckerman, I understand, who suggested that I might be given the task of looking at this whole complex of questions and report on it. So I became a Committee of one, with no previous experience in flood control or civil engineering or risk assessment to look at whether London needed a flood barrier. Again I was most fortunate in that my link with the Government machine was a delightful man, Colonel Stuart Gilbert, of the Ministry of Health, which in those days dealt with Central Government matters affecting Local Authorities. I consulted very widely, saw people from the Port of London Authority, London Transport, the local authorities, etc. Colonel Gilbert was always most helpful and of course he and I went to The Netherlands where their enormous expertise and the magnificent response to the disasters of 1953 had to be seen and understood. An incident then is perhaps worth telling: when, at the very nice dinner they gave us after the first day's work, they expressed their astonishment to Colonel Gilbert that it was he from the Ministry of Health who looked after this problem for the United Kingdom. "Don't worry," Stuart Gilbert replied, pointing at me, "he's an astronomer." I am sure our Dutch friends were confirmed in their view that the British were totally mad. I found it a most challenging problem, perhaps particularly so in looking at the question of what risks could or could not be accepted, what the meaning of probability was in relation to so great a disaster as one was envisaging. I came to the firm conclusion that Central London had to be protected, that it was not necessary to put a barrier far down the river, but that the Woolwich area would be the correct site, and wrote my report accordingly. It is often said that reports to Government, however well argued, just gather dust on the shelves. This was emphatically not the fate of my report, which convinced everybody concerned as soon as they read it, and within a year and a half (given the Parliamentary timetable, this is most remarkable), the Act for the London Flood Barrier was on the Statute Book. Of course I had nothing to do with the construction or the detailed design, though I enjoyed visiting it a couple of times, but the site was the one I had chosen, and the need for it was universally accepted. Through poor labour relations and other complexities, the Barrier cost far more than had originally been estimated. But even if the eventual cost had been known, I suspect that my proposal would have been accepted, but after long and heated arguments.

From that time also dates my involvement in defence. John Kendrew, my old friend, spent much time in the Ministry of Defence, with Solly Zuckerman. It is not often realized that John Kendrew did his outstanding work on molecular biology, which gained him the Nobel Prize, while he held a part time job in Defence. One day Solly said to John that it was desirable to bring fresh blood into the scientific side of defence, and my name came up. Apparently, so I am told, the conversation went that Solly said — "Well I am sure he would be good, but would he be interested?" John Kendrew, remembering my deep interest in military matters, replied with "Oh, he's the most bloodthirsty man I know." And so I became involved, first in the work of a committee to look at the vexed question of whether the country should invest in a new generation of aircraft carriers, ships of great cost and capability. Technically, it was a very important and interesting question of considerable difficulty, but overshadowing it all was the question of strategy. Was the chief British defence interest in NATO in Europe, or was it worldwide? For the second role aircraft carriers looked pretty indispensable. But was the country prepared to go in for this expense? Since the stunning disaster of Suez there had been several very successful and useful episodes. The support of the emerging country of Malaysia in its confrontation with Indonesia ended successfully with a total change of regime in Indonesia to one which had no such aggressive intentions. The support of the newly independent governments of East Africa against mutinies of their armed forces preserved order in Kenya and Tanzania and Uganda, the last one sadly only for a rather short period. The work was interesting and brought me into contact with many different strands of thinking in defence and in the government. But I do not think our committee influenced the world very much. But evidently my contribution as a lowly and new member had been enough for me to be asked, towards the end of 1964, to chair a committee to look at British defence interests in space. This turned out to be very important for my future, for it was my first contact with space, my first appreciation of what modern technology was like, my first involvement with international affairs. For, of course, we could learn a very great deal about space from the Americans. I reported to the Minister of State in Defence, Lord Shackleton, a wholly delightful man of great ability, whereas in the Committee it was Chris Hartley of the RAF, the son of a very distinguished scientist, with whom I worked with the greatest of pleasure. I believe the Report of the Committee is still not declassified, but I can say that what interested me most and where I put the weight of my recommendation, was in favour of small, inexpensive terminals on ships for satellite communication and the desirability of having, or participating in, a military communications satellite. Only tangentially involved in the issues of my Committee, but of relevance later, was that we formed a very negative opinion of ELDO, the attempt at an independent European launcher for space. Though purely civilian in intent it was sufficiently close to military affairs for each government to handle its own side of things with a very weak central organization. It was clear to us that such an endeavour could not succeed. Although it had been started by Britain as a way of making use of the successful development of Blue Streak, we were emphatic that this was not a good enterprise. That anything with an absurd management structure like this ever got as far as it did speaks volumes for the capability and devotion of the staff. But my sorrow is that it ever got going. Had the Government accepted my advice as soon as it was given, international complications would have been far less. As it is they only went that way in 1968 when I was Director General of ESRO, and there were serious international repercussions of Britain pulling out from a proposal in which other countries had already invested a great deal.

One episode from this period is perhaps also worth telling. When I took my Committee to Washington and received great help, advice and warmth from our American colleagues, I went to the Embassy and wondered whether we might not say "Thank you" with a cocktail party.

The Embassy immediately said "Yes", but, they said, it so happened that they had just had the same request from another British Defence Mission here under Rear Admiral Edward Ashmore. Would I mind it being a joint party? Of course I was happy with this and the invitations went out under our joint names. I met Edward Ashmore then and he became a good friend later, but one incident at that party was that an American came up to Edward with the invitation card and said, "Hermann Bondi, that is not a very English name is it?" Edward said, "No it isn't, my name is Edward Ashmore, a very English name, and my mother is Russian", which was quite true. It silenced the enquirer readily enough. Later on Edward and I worked together a great deal in Defence.

All this was very intriguing work for somebody who was no more than a Professor of Mathematics and not even a Head of Department, somebody who, only ten years earlier, had shied away from all administrative work. The report of my committee remained classified but was described in Parliament in glowing terms.

Another activity of the early and mid-sixties was to talk in public about science. One was a Grenada Lecture, which stressed the importance of communication for scientists, a recurring theme. I entitled it "Why Scientists Talk" and my friends said it should have been called "Why One Scientist Talks". I gave a few shortish radio talks, and then two other things happened. There was once again a food deal of noise about the so-called clock paradox of relativity, in which Dingle took part very much on what I thought the wrong side. Herbert Dingle, then Professor of the History and Philosophy of Science at University College, (who earlier had worked energetically in spectroscopy) and I had crossed swords before. Not long after the steady state theory came out, he, then President of the Royal Astronomical Society, gave a Presidential address which was largely concerned with attacking theoreticians in general, those propagating the steady state theory secondly, and me in particular. It was one of the most vituperative Presidential lectures ever given, and in my view complete nonsense, given on an occasion when no response could be made. Now, maybe seven or eight years later, Dingle again ventured forth with what was seen by me and indeed all physicists to be not only wrong, but dangerously wrong. For Dingle used his considerable powers of advocacy to tell the general public not only that physicists were wrong, but that we knowingly misused our standing to preserve a convenient, but, as he put it, inconsistent theory.

The Special Theory of Relativity, first put forward by Einstein in 1905 and the universal basis of all modern physics, always had a reputation among the general public of being obscure and contrary to common sense. I thought this was an absurd reputation for a theory that was bread and butter to physicists, and thus I was led into rethinking the foundations of a theory, in a way that perhaps had not been done since Einstein himself presented the theory. While others had constantly stressed its contrast to Newtonian theory, I presented it very much as its natural continuation and completion. This led first to an article in *Discovery*, then a very interesting request to do a series of articles for *The Illustrated London News*, with a delightful octogenarian illustrator and finally to TV work. There were two TV series in which I was the sole speaker, each, I think, of twelve programmes of about 25 minutes each. The discipline of television, the need for a pencil and paper man like me to use visually appealing experiments, the timing and the close links to those who set the experiments going, all this was quite novel to me, and a great challenge. I enjoyed it greatly and both series were turned into books, *The Universe at Large* for the series essentially devoted to astronomy, *Relativity and Common Sense* for the one on relativity. This latter book which was very widely sold and translated into numerous languages and was, 15 years later, reprinted by Dover in their "Classic Text Series". I conceitedly think that it is still the best account of the theory. However, this matter also made me think about the problem of improving presentations in science to make new insights easier to grasp for students and to popularize them to a wider audience. Such improvements require a great deal of hard work and hard thinking, but are highly desirable to widen and deepen understanding. Unhappily there is nothing in the career structure of scientists to encourage them to divert their time and energy from original research to work on improving the presentation of older achievements. While it may lead to mildly lucrative efforts, like publishing books or devising a television programme, it has little if any effect on promotion in a university or on being elected to the Royal Society. Do we need to change the system so that work on presentation receives sufficient priority? If, like myself, one has reached the top of the tree (Professorship and Fellowship of the Royal Society) reasonably early in life one can do such work as the liking takes one; but otherwise it is liable to have a negative effect on one's career.

Another event of those years was that Christine started teaching. Though she had been supervising in Cambridge, even after our first child was born, after the move and the growth of the family there was clearly no such possibility. In the early 1960s, though, we had very good live-in help in the house, and although the youngest child was only 3, she felt it possible to start a little part-time mathematics teaching which, over the years, grew to a very full-time commitment. Partly because of this, and partly because of my non-University activities, this period also saw a change in our economic fortunes. A Professor's salary plus the desire to have a good house in an excellent position, plus five children, did not make for an equation which was lush in any sense. So, for example, we had to be very economical with our cars. Our first one, which we bought in 1952, was a 1936 Rover 12, which we ran up to spring 1958, when we got a 1952

Super Snipe. During the week the car was of course always needed by Christine with the children, to take them to and from school, or to the playgroup, or whichever, so I had to walk or cycle to the station, when my travels did not coincide with the children's needs, or her other availability. Our holidays were usually in South Devon on a farm, very agreeable, but scarcely adventurous. Then in the early 1960s all this changed, and from that time on we really never had to think about money. We were never keen on flashy cars, and our children continued to be in the maintained system of schools which we found entirely satisfactory. Our only real luxury was that we loved to travel. In particular, in 1963, for the first time, we took the family on the Continent, down to the Costa Brava, with a long route through lovely places in France there and back. In 1964 Christine and I went with Cliff Price (Headmaster of the primary school) and his wife Lynne, also later head of a local school, to New England, disposing of the children in other ways. In 1965 we again went to the Continent, this time to the Brenta Dolomites. We do not like expensive and plush hotels. When we can, we like to take an apartment, finding this more comfortable and also of course cheaper.

CHAPTER 11

European Space Research Organization

IN THE spring of 1967, many of my endeavours reached fruition. The contract was signed for the first stage of the rebuilding of King's College; the State Treaty on the Anglo-Australian telescope was signed. My Report on the London Flood Barrier had been handed in, my work on the significance of space for defence was complete. I was not exactly short of work. There was plenty to be done, but I explain this to say how the circumstances made me so ready to jump at the opportunity that came. One weekend in early July Christine went to a conference in, I think, Glasgow. I drove her to the airport on the Friday evening and when I came home, Alison, our eldest daughter said "Daddy, somebody rang from the Department of Education and Science. He says it is very important and would you ring back, never mind what the time is." Then I was told that the European Space Research Organization (ESRO) which had its headquarters in Paris, needed a new Director General, that it was entirely appropriate that he should come from the United Kingdom, and could the Minister put my name forward for ratification by the other Ministers early the following week at their Ministerial meeting in Rome. It was not difficult for me to decide. I had to explain that my French was miserable, and that I proposed to go on living on Reigate Heath, because certainly there was no point in upsetting the childrens' schooling or my wife's career. And so I gave these as the two, and only two constraints over the telephone, saying of course I wanted to confirm it with my wife when she came back, which was on the Sunday night. Before long I went to Paris and was formally appointed to the post. Of course, there were difficulties for the College, and my initial reaction had also been to say that until Christmas I could only be part-time, but fortunately my colleagues made the necessary arrangements and I could take over from the 1st of October. This was of course an entirely new challenge for me, and I must confess I did not realize how big a challenge it was going to be. After what I had done I was full of self-confidence, but there were many moments in ESRO when I felt stretched to breaking point. This was responsibility of a far greater order than I had ever experienced before, although when I had said "Yes", I had not totally appreciated this. I think I can honestly say that I had never enjoyed a job more than ESRO, particularly perhaps because it was so challenging. Originally I took the job on three years leave of absence from my post at King's College, and had every intention of going back there. I liked the people, I liked the students, I liked the teaching. And so an appointment of three years seemed quite reasonable to me. The mode of living, it turned out, suited me very well. I had a little flat in Paris, where in fact I only averaged two or, at most three nights a week. The job involved a great deal of travel to ESRO's other locations, to capitals for discussion with governments, to visits to industries, and to contacts with our friends and colleagues in the United States. But it is my proud boast that in the three and a half years I eventually was with ESRO, I never missed getting to my family on Friday night, unless I was in another Continent which happened maybe two or three times. Usually this meant getting home to Reigate, but we also got together sometimes in Paris or elsewhere. Of course it meant an early Monday start, but it did mean that during the week I was totally and uninterruptedly on the job, able to concentrate on it, with nothing else in my mind. And that, I am certain, is the best way to manage anything. The flight home on the Friday evenings was also excellent, in that after a heavy week it brought the shutters down over the work, and the weekend was entirely devoted to the family.

When I paid a courtesy call on the Minister for Higher Education, Shirley Williams, and explained to her how I was going to live, she said that poor ESRO always had its crises at weekends. "Well," I said, with supreme self-confidence, "when I'm Director General they will not be at weekends." And they were not at weekends in my three and a half years, severe as they were.

I took over from a great man, Pierre Auger, the first Director General. A wonderful scientist and great internationalist, who had founded both CERN and ESRO, but with all his great qualities management was not his strength, and in many ways I was free to create my post in my own image. The driving force in ESRO's early history had been Freddy Lines, of the SRC, who, as Technical Director, had driven forward energetically, successfully and ruthlessly, stepping on many toes on the way. I reaped what he had sown, mainly good, but in some diplomatic aspects in particular I had to be very gentle and understanding of the deep wounds that were felt. His tremendous drive and energy had led to ESRO, devoted at that stage entirely to scientific satellites, having already in such a short time completed a satellite with work far advanced on two more. Unhappily in May 1967 the American launch of the finished satellite, ESRO II, was unsuccessful and it had dropped into the Pacific. Freddy Lines was leaving too. In fact there had been a complete re-organization that greatly enhanced the powers of the Director General and the Duirectors *vis-à-vis* the member states, and I got an almost entirely new team of four Directors in the selection of three of whom I had a hand. All four became very close personal friends, and many years later we still see each other from time to time. The only survivor from the previous régime was Marcel Depasse, a Belgian diplomat who did all the difficult work of keeping the organization together, and who was a special help for me in teaching me the rudiments of diplomacy. We had ten member states. The organization was wholly separate from the European Community to which, of course, the United Kingdom did not belong at the time anyway. We had France, the Federal Republic of Germany, the United Kingdom, Italy, Switzerland, Sweden, Denmark, Netherlands, Belgium and Spain. Each of the member countries had its own interests, had its own complaints, had its own concerns. My next Director to join the team was Dinkespiler, a superb French engineer, who had been very active in the French National Space Organization and who had been a member of the committee that reorganized ESRO. Depasse, Dinkespiler and I had our offices in the headquarters in Paris, and were in constant, hourly, contact, and the three of us worked together very effectively. ESRO's biggest establishment, ESTEC, the technical centre, was at Nordwijk in The Netherlands and as its Director we secured Kleen, a German engineer from Siemens, a little older than the rest of us, who kept an establishment of prima donnas on the rails in a most remarkable way. Finally, ESRO was just beginning to establish an operations centre in Darmstadt in Germany, for which I secured the services of Umberto Montalenti, an Italian engineer of great distinction. The three Paris directors and the two directors with responsibilities for large establishments met frequently in long meetings, to sort out our many problems, most of which of course were people problems. I learned that however difficult the technology, if you can resolve the management problems and keep the people working together, then the technical problems will go away. I profoundly believe that technical failures, due to technical difficulties, are very rare, while almost every failure can be traced back to management defects.

I will mention briefly a few of our problems and how we overcame them. Though one usually builds two models of a satellite at the same time, the second model of the satellite that had fallen into the Pacific was by no means complete: because, when there are two models, the one being prepared for flight always has precedence: bits and pieces out of the other one are used, without much caring about what is left behind. So to get the second model into a reasonable shape was now a major task, as became clear to me at the very first meeting I went to, even before I fully joined the organization. I immediately saw that our reputation would be mud if we could not get the second flight ready at reasonable speed and with perfect reliability, and that that would need exceptional measures. I therefore appointed our finest engineer, a Frenchman by the name of

Blassel at ESTEC, to take the co-ordinating role in this and make sure that this satellite prospered, as indeed it did. Blassel was my saving and my headache during most of this time. An absolutely outstanding engineer, a perfectionist, as one has to be in space engineering, he was deeply incommoded privately and therefore disgruntled for being in ESTEC rather than in Paris. Like every good perfectionist engineer he always wanted more resources than he could have, which led to difficulties, and a good deal of my time and energy was spent in keeping him and the rest of the team working together. But that was time well spent. The next problem that arose was far more serious. After its first three satellites, all of modest size and complexity, ESRO had embarked on a pair of large ones, and these had been accepted by the Council on a cost estimate that turned out to be grossly wrong. When I and my team began to understand the situation a great deal of money had already been spent. Moreover, the award of the contracts had left one of our major member countries, Italy, feeling deeply disgruntled. The amounts involved would have rendered the future existence of the organization questionable, to put it in its lowest terms. But behind all this there were another two quarrels between member states. One was that although the constitution of ESRO said that the member states agreed on a three-year level of resources, this had been breached, essentially through the action of Belgium, which had been very suspicious of various features of the then management. Thus we lived from hand to mouth, and effectively illegally. Secondly, at the same time as the ESRO Convention had been signed, the United Kingdom had proposed and other states had agreed to set up a European Launcher Development Organization, to build a big launcher. This organization, ELDO, was in a far more difficult position than we were. For the countries concerned, which were the United Kingdom, Federal Republic of Germany, France, Netherlands, Belgium and Italy, regarded launchers as related to military matters and so did not entrust the central organization with any great power. Each dealt with their own industries, but had agreed on a sharing out of the parts of the launcher and on the percentages of contribution. So if, as happens so often in technology, one of the cost estimates turned out to be too low, the others felt cheated if the items their industries produced did not rise in the same manner. It was a superb machine for cost escalation, and it did its job. In my Defence Space Committee in the United Kingdom, markedly earlier on, I had said that ELDO did not make sense and should be cancelled. I am sure it was the right recommendation at the time. Modern technology cannot be built in such a piecemeal fashion, but unhappily it took two and a half years before this cancellation became official Government policy, the other countries had already invested substantial sums and therefore did not want to agree to this, and so there was a terrible row which was in the background all the time during my stewardship. ELDO never succeeded in putting an object into orbit. That it came rather near to it, in spite of its managerially impossible set-up, is a great tribute to the people who worked for it. Only later was ELDO finally cancelled, and succeeded by the tightly managed Ariane programme.

It was against this illegality, lack of assurance of the future, deep divisions between the member countries, that we had to deal with the pair of satellites whose costs had risen so alarmingly. Italy, whose very competent industry had been totally neglected in the award of this contract, did not want to contribute any further to this satellite pair, and we did not feel the Organization could afford more than one. So we came out with the solution that the other member states would pay only for the completion of one of the satellites (TDI), and all would then contribute towards its operation. It stretched our legal people to get this right, but we managed this. Two further of the small satellites which our predecessors had started were completed and launched in 1968, and operated well. By the time the space ministers of Europe met in Germany in autumn, ESRO had solved its biggest technical and political problems with the big satellite (TDI), had three well functioning satellites in orbit and looked set for big things. In view of these successes, the ministers of our member states agreed now on the three-year level of resources as the State Treaty demanded. We were now legal, with assured finances and things went very well. Soon afterwards, the United Kingdom defaulted on ELDO, which made the background most uncomfortable. And we had many more crises, in part through involvement with American projects, in part in other ways. We started several other satellites all of which came to fruition successfully. However deep political divisions, it is difficult to kill an organization that is technically excellent and successful. But there were quite a number of occasions when I did not know, from one day to the next, whether we would still exist. Just keeping an organization of this size going (we were about 1200 persons) takes a lot of nervous energy, management and sheer hard work, as I discovered.

Perhaps a few other points should be mentioned. ESRO had been founded by scientists to foster European science and technology, but the science was thought to be entirely in the driving seat. My scientist friends determined which satellite we should build for scientific purposes. Yet it was becoming clear that Europe was also going to need application satellites, first in telecommunications, then in meteorology, then perhaps in remote sensing and, one application that I was particularly keen on, aeronautical satellites for constant communication and position determination of aircraft, a project that never succeeded in being accepted. Since ESRO had been founded by scientists to build and operate scientific satellites for their objective, scientific research, we first had to persuade our scientific friends in the member countries and through them their governments to widen our remit to include applications satellites. Then we had to convince our member governments that if they wanted applications as well as science from the organization, they would have to pay more. And, third, we had to persuade the member countries with national organizations, that they should leave these applications largely to the European Organization rather than do it themselves. This took a great deal of effort, but we got there. With the Italian situation fully settled, our biggest problem was, perhaps, with Spain. Of course we were still deep in the Franco era, with Spain practically an outcast. It was at the insistence of the Spanish Foreign Ministry that Spain joined, but this was never liked by those in Spanish industry and in the airforce, who felt that they were most unlikely to get any contracts, so that Spain, as they put it, was essentially paying into ESRO for the benefit of the more advanced countries of Europe. This led to the real threat of a Spanish withdrawal and the upsetting of the whole mission. What was needed was, first, diplomacy to make clear how highly we valued Spanish membership, and secondly, real action to get contracts to Spanish industry. For this purpose I sent a mission to Spain to look at the potential of their industry and see where the competence and equipment was high enough to take part in these space ventures. In spite of a certain inferiority complex on the part of much of Spanish industry, we found plenty, and had no great difficulty in steering sufficient contracts to Spain to make them quite content.

I must confess I had my own political worries there. I had been young enough at the time of the Civil War to feel real dislike for the Franco regime. Yet, there it was, and I was hobnobbing with its Ministers, and helping it along. I was clear that I was paid to be a good European, not to indulge in my own political prejudices. At the back of my mind there was the belief that prosperity is a great help in making countries more reasonable in their politics. How well we saw that belief come to fruit in Spain! Some years after my time Spain became democratic, contributing to the triumph of democracy through the width and length of Europe. I had seen some such progress earlier in the old South of the United States. When we had driven through it in those very early days of 1954, one accepted the curious local habits and one certainly did not feel like meddling in a foreign country and so alien a part as the old South. But it was all extremely poor, and what we would now call a developing country. The next time when we were there in 1960, the situation was changing. There was much more industry and wellbeing, but restaurants in general were still segregated. We almost universally went to take-away places, which were the only ones where no segregation by race was in order. When we went again to those parts, about ten years later, segregation had been totally forgotten, and it was one of the most prosperous parts of the United States. Prosperity is a cure for many ills, which is why the most reactionary and racist groups in South Africa have always been against industrialization, against business, against prosperity.

One or two travel stories of that period of my life may also be worth telling. Very early in my ESRO period I was very keen to visit Darmstadt and its new operations centre, and meet the people there. But I was horribly busy in my Paris office, so arranged to take the lunchtime flight to Frankfurt, where they would meet me, spend the afternoon in the lab., the evening dining with the senior members of the team, and take the sleeper back from Frankfurt to Paris, so missing only half a day in the office. But the plane was diverted when it got over Frankfurt and after circling for some time, landed in Cologne. After landing we were told we would be taken by coach. After further delay, the information came that Frankfurt was clearing and so we would be flown. When this, too, came to nothing my patience went and I rented myself a car to drive down. It was a nightmare drive. In the hills the fog was thick, the autobahn was under repair for most of the way, with very narrow lanes and an unending procession of lorries hurtling past me at great speed, with occasional really bad holdups. The upshot was that I got to Frankfurt in good time to catch my sleeper back to Paris, but I had never seen anybody at all.

More successful was a visit to India. The Indian space effort, just beginning under Vikram Sarabhai, was very keen that I should pay them a visit. I have since been back several times to look at the flowering of this great enterprise. But I was under great time pressure. And so it worked out that one week I flew to Paris as usual on the Monday morning, and had all day in the office in Paris, that I came home as usual on Friday night, having had all Friday in my office in Paris, but in between I was in India, spending a good deal of time with the space people in Delhi and in Ahmedabad, and also had time to see the Taj Mahal more or less on the way.

A third story came to me from one of my senior colleagues who, like me, had to do an enormous amount of flying in this widespread organization. As was very common in those days, one day he was asked at an airport to give details for a survey of travelling habits for the Government. This young lady had a long form on which she filled in all his replies with great care and precision. Near the bottom of the long form came the question, "How many times a year do you fly?" My colleague answered, quite correctly, around 150, maybe 200 times. The lady turned stone cold, said "Thank you very much" and as he walked away he glanced back, and saw that she had screwed up his form and thrown it away. Incapable of believing that there were people like that meant that she would not accept his information and no doubt the Government could then say, "We have had a complete survey made, there are no people who fly that often."

CHAPTER 12

Defence

As THE summer of 1970 approached, with my appointment due to end at the end of September, the question of my staying on came up. I was quite keen to stay a little longer. The Council of ESRO seemed happy for me to stay on, and so both sides said, "Yes", with no very definite end date in mind. I was looking forward to my return to academic life, though when one has been used to a wonderful private office, with two secretaries and a Staff Officer, to a car and driver, and all that which makes life easy and makes one so much more effective, there was just a little worry at the back of my mind about the austere circumstances of academic life. But this was not at all in the foreground of my mind when Solly Zuckerman asked me to come and have lunch with him, and told me that he had talked to the newly appointed Secretary for Defence, Lord Carrington (whom I had briefly met many years earlier in Australia), and they both were most anxious that I should become Chief Scientific Adviser to the Ministry of Defence. This proposal came as a total surprise to me. Much as I had been interested in Defence, to make it a full time occupation had not crossed my mind. Moreover, heading ESRO, an organization which had two neutrals (Sweden and Switzerland) amongst its members, I had steered totally clear of anything defence-related for three years. And it took me a little while to make up my mind, but Solly's urging, Peter Carrington's charm, the likelihood to meet old friends again in Defence, all made me decide to say "Yes". But there was my obligation to ESRO, which got into one of its bad crises shortly afterwards in the autumn of 1970, where I just could not leave them, particularly as the succession was in no way decided. But gradually things calmed down, and I took my post in Defence on the 1st of March 1971, having said goodbye at the end of the previous week to all my friends and colleagues who had given me so memorable and enjoyable a time in ESRO.

Perhaps this is the place where I should describe my attitude to defence and to the armed forces. First, it is clear to me that no task of government is more important than to keep the country safe from suffering a major war. All the other tasks of government would be null and void if such a disaster were to befall us. There may be legitimate differences of opinion about how best to achieve this overriding objective of minimizing the risk of war; there can be none about the importance of the task. Second, the well justified horror of the use of nuclear weapons must not lead anyone to regard war with conventional weapons as anything but a total and complete disaster. The most fiendish consequence of war has generally been famine. In the First and Second World Wars this was narrowly avoided in this country (though not in the Soviet Union, especially in Leningrad, in The Netherlands, in Bengal). Modern agriculture while highly productive, is extremely vulnerable. If the electricity fails the cows cannot be milked, and the stores of food would go bad. Without a supply of fertilizers and pesticides, the yield of farming would diminish drastically. Can one conceive of modern conventional warfare in which electricity generation and distribution would not be prime targets, as would transport centres? All that we now regard as essential to civilized living, like hot water, space heating, public and private transport, telecommunications, medical facilities, etc. would be drastically reduced in availability. With food scarce and food hygiene much reduced, epidemics would be severe. Even in the far less vulnerable society of seventy years ago, the aftermath of the First World War led to an influenza epidemic that killed many more people than the bullets had done. So the task of government that is so overwhelmingly important is the avoidance of war, not just of nuclear war. Unless this essential point is understood, a rational strategy for survival cannot be discussed.

Rightly, we belong to an alliance, NATO, binding together almost all the countries on whom we most depend and to whom we are a vital partner. If anyone of us was attacked, we would all suffer, so we must stand together. In the past, major powers have frequently engaged in wars of "limited liability", wars (such as colonial wars) in which success could be advantageous, but failure would not in any way threaten the existence or the vital interests of the aggressor. How tempting Central Europe must have looked all those decades to the USSR as an area to stage a conventional war of "limited liability"! Not much would have been needed to stop the, to them, not just irritating but threatening growth of the West German economy. The, to them, painful existence of the contrast between well being and welfare in Western Europe and the miserable showing of the East European satellites could have been abolished at a stroke. And the risk to them of such a military adventure would have been slight, for the idea of NATO armies, armed conventionally, advancing into the Soviet Union was effectively zero. The one thing that stopped any such temptation in its tracks was a nuclearly armed NATO, for in the nuclear age there are no wars of limited liability. I know of no other period of history before our time where many decades of extreme tension have passed without any military action. The logic of ascribing the preservation of peace in Western Europe to the existence of a nuclearly armed NATO seems to me incontrovertible. Deterrence has been the key to our security.

Yet more than nuclear weapons are needed to preserve peace in Europe. Too great a conventional imbalance against us would make the nuclear trigger too sensitive to be credible. So our contribution to the conventional strength of NATO seems to me to be equally essential. Thus I completely support British membership of NATO, with its doctrine of conventional resistance to aggression, with a nuclear backup available, because I believe it to be the key to the preservation of peace which is so absolutely essential.

Compared with these fundamental points, Britain's own nuclear capability is a matter of balancing advantages and disadvantages. First but least discussed, is Britain's nuclear capability in anti-submarine warfare. For geographical and historical reasons, the United Kingdom's contribution to the alliance's task to ensure the security of the Atlantic lifeline is pre-eminent amongst European partners. The greatest threat to this vital link is posed by the putative opponent's nuclear propelled submarines. They are so thick skinned that a torpedo with a conventional warhead cannot present too great a danger to them. Thus if Britain is to make a contribution to the security of the Atlantic it must have its own nuclear warheads, since on a ship, aeroplane or submarine warheads owned and guarded by the United States (as is a well established practice on land) not feasible. Thus at sea, a denuclearized Britain is a toothless Britain.

The question of a strategic deterrent, on which so much of the discussion centres, seems to me more finely balanced since in some respects it adds to our security while in others it detracts from it. But in the final analysis, if the cost is tolerable, it is advantageous to have the country in a position where nobody can ever risk driving it so much into a corner that it could go crazy.

Some people may feel that whenever there is a thaw in the cold war, we should throw away our weapons. But this ignores the fact that military intentions can change much more quickly than military capabilities.

Others may think that our security would be better served by neutrality than by membership of NATO. For a country of the United Kingdom's size and geographical position this is indeed an option, provided the country's strength is made great enough. Forces based on two year conscription, with perhaps twice the number of weapons now held, might be sufficient. Do the advocates of neutrality feel they would like to face the bill?

Thus I wholly support the line on defence that government after government has held, and had no difficulty in serving in a Ministry of

Defence that regarded the preservation of peace as by far its highest priority.

There is another criticism that one sometimes hears: should scientists be prepared to work on fiendish weapons? Not to do so would have required all scientists to combine in agreeing to refuse to do certain types of work for our political masters. To me, a democrat through and through, such a conspiracy of scientists against the legitimate elected government is no less abhorrent than a military conspiracy against the legitimate elected government. I would never wish to serve an authoritarian unelected government, however peaceful its intentions. With a democratically elected legitimate government, decisions on these morally and ethically immensely difficult questions are for ministers, and not for me. To explain myself clearly, to describe potential horrific consequences with all the lucidity that I can muster, that is my task but the decisions belong to those entrusted with the power of office by the people.

Yet one must recognize that there are limits, limits that never seemed to be anywhere near my area of work. While objecting to scientists combining to subvert a decision of government, I fully uphold the right and indeed duty of individual scientists to decide, in the light of moral and ethical considerations, whether or not they are prepared to work in a particular field.

Indeed the central point of my attitude is that citizens should *think* about problems of national security. I entirely accept that different people may come to different conclusions. While I respect those that differ from me, I am bothered by the many who do not regard national security as an issue one should concern oneself with.

Perhaps I can illuminate my attitude with a little story. My colleagues in the RAF asked me to be the after-dinner speaker for newly qualified navigating officers. When the time came for me to speak in the RAF mess, there were all these young men sitting in front of me. Having worked hard for their exams and having imbibed a great deal of liquid refreshment with their dinner, there was every risk that they would go to sleep as soon as I started talking. So I began "You are just like the Aldermaston Marchers. They and you think hard about defence. It is true that they come to different conclusions, but this common concern sets them and you jointly apart from so many people who do not think about these matters." Nobody went to sleep!

My strongly held attitudes did not mean that I found working in defence easy. To have to think about desperate situations is a strain, but to feel that in however small a way one contributes to the preservation of peace makes it all worthwhile.

In spite of my previous contact with Defence, my new existence was entirely novel to me. I was now a Civil Servant of the highest rank, that of Permanent Secretary, on a level with the Chiefs of Staff of the individual Services, and only marginally below the Chief of the Defence Staff and the Permanent Secretary of the Ministry. I was dealing with Defence, not as an abstract exercise, not even in the areas of minor actions of interventions all round the world, but on the central matter of the European confrontation, which was now *the* chief issue in Defence. One had to think all the time about the worst that can possibly happen, a major war in Europe. All one's endeavours were directed to avoiding this, yet one *had* to avoid it by showing that one was ready to resist and respond, should the need arise. So one had to think about very ugly situations indeed. It took me some time to become accustomed to this. It was again a tremendous challenge, but one very different from ESRO. International links there were in plenty, with the United States, with Germany, with France, with Italy, with, indeed, all our NATO allies, but emphatically the Ministry of Defence was not *my* shop. In ESRO I could truthfully say, as the boss, that if something went wrong in the organization I did not have to ask whose fault it was, it was mine. In Defence, however elevated my rank, I was just one cog in a big machine, in a machine of such importance that, very rightly, its head is a politician, a member of the Cabinet, one entrusted by the popular will to take the most difficult decisions there can be. Again I had all the trappings of office. Of course I could not ask King's College to keep my post open any longer, and so I had to resign my academic position. In fact the University of London and the College, in their kindness, went through a procedure that again made me a Professor, albeit in title only, which I held until, in 1985, even that ended due to my advancing years, and I became Emeritus Professor.

In the New Years Honours of 1973 I became a K.C.B. (Knight Commander of Bath) which is a normal honour for a civil servant of my rank, though mine came rather speedily. Though at first I was not too sure about my attitude to it, this was soon settled when I saw the pleasure it gave to others and the added weight it gave to my actions. This emerged first on my mission to India in January 1973.

The challenge in Defence was wonderful. The system was superb in making me and my small group independent of the chief procurement agencies, the Government Research Establishments, and the operational requirement staffs of the individual forces. So when projects had reached the stage where a Yes or No was required for their development, before large funds could be spent, they had to come before the Defence Equipment Policy Committee which I chaired and greatly influenced. In this position therefore I was in no way tied by previous discussions. I and my staff could look at the issues in an entirely independent manner and come to our own technical conclusions, ask our own searching questions. And so I learnt a great deal about new technologies and, as had already happened in ESRO, I began to acquire some feel when to trust the evidence on a proposed development as regards estimates of time, performance, and cost, and when not to trust it. Of course, an outsider can never pit his feel against the expertise of those who devised the whole scheme, even less against those who can speak of its military advantages and disadvantages. But one can put in awkward questions and, in particular, one can put in milestones. One insists that the most critical parts are studied first, are worked on with relatively minor expenditure, and the next slice is not released until that milestone is passed, at the required time, at the required cost, with the required performance. I do not want to say that we always took the right decisions, but I do think they were markedly better because of the existence of the Central Scientific Department, and myself as its Head.

I had my great worries though. In ESRO and in all the industries I had been in contact with, a project was managed firmly by a Project Manager with great authority, a person who saw the management of this project and its success as the make or break of his career. He stayed with the project from the beginning to completion, and so at every point this Project Manager knew that he only had to deal with problems he had himself created, which is a tremendous spur to overcoming difficulties. In the Defence world, normally it was a serving Officer who was a Project Manager. The need to gain wide career experience in Defence means that these people are moved about every two and a half years or so. So they saw this as just a piece of their career. The new person coming in would not feel the same commitment to the choices that had been made before, some of which now appeared to have been rather ill-advised. Accordingly there was a going over the ground again. It is rare indeed that a project can be completed in the tenure of one person, if every two and a half years there is a change of Project Manager. Each change can add a significant amount of time, perhaps up to a year to the length of the project, so the long projects become very long indeed. This can mean that they may no longer be adequate to meet the threat of the future that has meanwhile developed. So military procurement faces problems of its own kind, which are very difficult indeed. Again, as I have already hinted, one is always measuring one's success against a moving target. Making an anti-tank missile may be suitable for one kind of tank, but not if a new kind, with very different armour, appears. And so the danger of a project becoming irrelevant is always with one.

Industry sees Defence as a peculiar customer. Good in that they stay with you through thick and thin. Good in that the customer must pay the price, whatever it turns out to be. Bad in not preparing the industry concerned for the civil commercial market, bad in not giving sufficient priority, in industry's view, to making defence products saleable to other countries. Yet there is no easy way round this. Competition is of doubtful value because, at best, there are very few companies able to develop say, missiles. If you suppose there are three, and after competition the first missile project goes to one of the companies, then maybe the second one goes to another company. But when the third project comes up, the third company will say, "Unless I get this one I will not stay in this business. I cannot keep highly qualified teams together for business that never arrives." And so you are just as much a slave of the third one in this case as you would be the slave of a monopolistic supplier. When there is no chance of large scale competition, then indeed competition may be a lure.

The sheer variety of development projects I had to deal with was great and most fascinating. There was constant contact with advanced industry and tremendous support from the rest of the central scientific staffs and many others. Of course, the Ministry of Defence is an enormous bureaucracy and it is very easy in such a situation for separate empires to arise, with attitudes that differ for reasons good or bad. The essential thing always is for the utmost honesty and frankness. If the people on top work well together, then this spreads through these large groups of people and leads to results which are more rational and better than they might otherwise be. I was most fortunate with my colleagues, who were people of outstanding ability. Through personal contact and the warm appreciation of each other, we managed to make the whole thing work as well as one can expect of an outfit the size of Defence. Naturally, the most important person in the place is the Secretary of State. To work under Lord Carrington was always a delight, with his sharp acute questioning, his constant goodwill, his understanding for and of everybody, and his power of decision. As I had learnt in ESRO, decisions generally are not exactly of the kind that this or that group or interest wanted, but decisions are loyally accepted if, first, there has been a good hearing for the opposing point of view and, secondly, there is real appreciation and admiration for the person making that decision. This was always the case in the Ministry of Defence where I served under a succession of outstanding persons. First, Lord Carrington, then a brief interregnum during Elections, when Ian Gilmour, a delightful man, was as much in charge as any politician is during an election period. Next, Roy Mason, a wonderful man for whom I have the utmost admiration, particularly for his power of decision making. And last in my time, Fred Mulley, with a razor-sharp mind and a personality that spreads good working relations all round. Similarly we had outstanding people as Permanent Secretary of the Ministry and a most interesting succession of Chiefs of Staff of the single services and Chief of the Defence Staff. Many became personal friends (for example Sir Michael Carver) and it is a pleasure to see them, a decade later, from time to time. The problems were never easy. Did one trust the promoters of a project, or should one buy abroad? Or get the design from abroad and build under licence in this country? These questions were all the time difficult to decide, and milestones were a great help, but even so not everything came up to expectations. From time to time, painfully, there had to be a cancellation of a project, knowing that one had wasted not only a great deal of public money, but the time and the talent, the ability and the enthusiasm of excellent engineers and scientists and service officers.

Perhaps the most difficult problem I had to deal with was the future of the British strategic nuclear deterrent, Polaris. I have explained earlier my views on nuclear deterrence and do not intend to repeat these here, except to stress the desirability for a British deterrent and the fact that the Polaris system worked well. However the development of Soviet ballistic missile defences around Moscow made its capability suspect, pleased though we were about the Anti-ballistic Missile Treaty, which effectively confined the spread of such systems. The Americans had changed to a much more complex weapon, Poseidon, which had ample penetration power, where our Polaris did not.

There were, then, three options (apart from giving up the British strategic deterrent altogether, a solution not welcome to the government). One was to do nothing at all and admit that the capability of Polaris against Moscow was doubtful, to put it at its lowest. The second was to improve and harden Polaris, a huge piece of engineering, particularly of space engineering, plus some engineering in the nuclear field itself, engineering which would have to be largely or wholly British. The third option was to persuade our American friends to let us have Poseidon as they had let us have the Polaris missile. It was common ground that the first option was most undesirable. Our deterrent was bound to be small in quantity. To have its quality debatable too, rendered its whole purpose doubtful. On the second there was a sharp difference in perception between the Royal Navy and my own group including AWRE Aldermaston. The attitudes of both sides were, I think, entirely understandable. My naval colleagues wanted above all a well-tried and tested system for their sailors to handle. The vastly greater scale of the American endeavour meant that an American system had very much more experience in it than a purely British one. It is very natural that those who are responsible for the people who man the submarines, and work in difficult and at times dangerous conditions should want to give them a weapon in which they can have the utmost confidence. So my naval colleagues favoured the purchase of Poseidon. The contrary view held by the engineers in my section and those at AWRE related principally to arguments involving the desirability of the British deterrent being recognizably different from the American one. The first point concerned the hot debate of those days about MIRV (multiple independently targetable re-entry vehicles), which was supposed to be a characteristic of undesirable nuclear escalation. The United States Poseidon was MIRVed, the proposed modification of Polaris was not. Moreover this lack of

MIRVing underlined the second strike characteristic of the British weapon. ("Second strike" means that it could be used as a response to a nuclear attack on the United Kingdom, and not as part of a pre-emptive attack on the opponent's nuclear strike force.) Next, an independently developed front end of our deterrent would strengthen, preserve and demonstrate the British capability both in space engineering and in nuclear warhead design and production. There were additional points, such as the desirability to minimize expenditure outside the country and the feeling that it might be easier to keep the British deterrent out of U.S.–U.S.S.R. negotiations on strategic weapons if it differed more rather than less from the U.S. system.

Those were the debates which were going on in my day, until eventually a decision was taken to go for Chevaline, the British improvement to Polaris. Until that decision was taken the project was kept in a kind of suspended animation, with funding six months by six months, with a total question mark over its future. In such conditions of indecisiveness one cannot transfer many people of the highest calibre onto a project of doubtful future, for there are other things for them to do where one can be certain that their abilities will have a positive effect. Nor is it possible to motivate people to want to go there. All this changed once the decision in favour of Chevaline had been taken. Edward Ashmore, as First Sea Lord, and I could advise the Secretary of State to set up a proper unified and powerful management structure. Chevaline illustrated a permanent problem of the political/technical interface. One cannot give the cost of a development project until it has been studied thoroughly by very good people for some time. Such a study costs a lot of money and uses the time of very able people. If one is not prepared to put in such an effort one cannot expect a good answer as to the likely cost. Once a project is firmly decided on and a powerful management system is in place, able people have been attracted to it and allowed to get on with it, then, and only then can one make a good guess at how long the enterprise will take and how much it will cost.

Though my work load at the Ministry of Defence was heavy, with a good deal of travel, I managed to find time for some professional activities. Every now and then I got round to doing a little research and so published a paper, but I became also very much involved with the Institute of Mathematics and its Applications. The Institute, founded on the initiative of my friend James Lighthill in the mid-sixties, aims to link and emphasize the mathematical content of the work of mathematicians wherever and however they are employed. This emphasis on mathematics as a tool for a very wide variety of purposes is wholly congenial to me. I was its President in 1974 and 75 and have been active in it ever since, speaking at local meetings and at conferences, writing for it, etc. Its first Secretary and Registrar, Norman Clarke, and the second one, Catherine

Richards, are both personal friends whose energy and drive has given the Institute its wide membership, professional standing and major publishing position which has made it most effective and influential.

Christine is also a member in her own right, has served on its Council at times, and is generally active in it. In 1988 the Institute honoured me with its Gold Medal inscribed to my great pleasure "Awarded for his brilliant applications of mathematics to the Cosmos, and to his signal service to the British nation."

My successor as President was the Duke of Edinburgh (which led to a number of memorable and enjoyable occasions) and his successor was Dame Kathleen Ollerenshaw who, like her husband, became a great personal friend of ours. She is a mathematician, educationalist, and deeply involved in local government in her City of Manchester whose Lord Mayor she was, as well as long-serving Councillor.

It may be worth mentioning here that somewhat later she stimulated me to work with her on a mathematical subject far from my usual fields, namely Magic Squares, and we published jointly a major paper on it in the Philosophical Transactions of the Royal Society in 1982. She has always been a person of immense energy, wholly undiminished by age.

During my time at Defence I received Honorary Degrees from the Universities of Surrey, Sussex and Bath. (Later such honours came to me from York, Southampton, Salford, Birmingham and St. Andrews, as well as an Honorary Fellowship of the Institute of Electrical Engineering and a Companionship of the Operational Research Society.) Such honours always give me a warm glow in the stomach and a very agreeable day.

Shortly after I started at Defence I became a Governor of the Ditchley Foundation, which organizes Anglo-American conferences at its most beautiful seat north of Oxford. We have been to many of its annual lectures and I have chaired one and attended several of its excellent conferences.

CHAPTER 13

Energy

MY TIME in Defence was certainly a period when I felt I was working at full stretch, when I fully entered the, to me, new world of the Armed Forces, made many friendships, and learned a great deal about different walks of life and different attitudes. Though my original contract with the Ministry had been for five years, I had had to cut my links with the University, as previously stated, and after a little discussion and negotiation I became a permanent Civil Servant in early 1976 instead of a temporary one as I had been hitherto. Not only did this regularize my position but it led to distinctly improved pension arrangements. Of course it was all only for a limited time. Sixty is the retirement age in the Civil Service and though exceptions are made at lower levels, they are not common at that of Permanent Secretary, the rank which I held, and so this would lead to my retirement in the autumn of 1979.

Before this other events intervened. Energy had long been an interest of mine, particularly as I felt that in the heating of homes great improvement could be made relatively easily. I had already put double windows into our house in Cambridge in 1949, and given it a primitive but tolerable central heating system where none existed before. This was far earlier than most people in this country did so, and I felt that it would not only add greatly to comfort but would actually save energy and therefore money. I had followed with great interest the moves that were being made following the oil crisis in the winter of 1973/74. And then one day, in the spring of 1977, I got an invitation from the Secretary of State for Energy, Tony Benn, to have lunch with him. This was not my first contact with Tony Benn. We had first met many years earlier when one of our friends in the Reigate area had asked us to come to her Literary Society meeting, where a young M.P. was to talk about the work of Parliament, and she was keen to have a large audience. Out of friendship we came along but we did not expect very much, because we thought an M.P. talking about the work of Parliament could only be either extremely partisan or as dull as dishwater. The M.P. was Tony Benn. He had us spellbound for an hour, and he never said a word that could be considered partisan. And so my appreciation of him began then. Next, in autumn of 1968, when I was Director General of ESRO, there was a ministerial conference in Germany. Tony Benn, then Minister for Technology, led the British delegation to this meeting which was presided over by Stoltenberg, the German Minister for Research. I had made every effort to meet Stoltenberg before the meeting but he, of course, was immensely busy, and all that could be managed was a chat during a car journey. But that was quite enough to lead to a tone and conduct of the meeting that suited our achievement in ESRO extremely well, with three successful satellites in orbit at the time. But the ELDO crisis, to which I have previously referred, cast a severe shadow over the conference. In the discussions Tony Benn, as always a little impatient with his civil servants, put forward an insufficiently thought through proposal, which I was able to shoot down in front of all his ministerial colleagues as he acknowledged. A lesser man might well have taken a dislike to me as a result of this, but Tony immediately became keen to have my help and in 1977 the chance came, because after profound disagreement with Walter Marshall, who had been his Chief Scientist, he had a vacancy and offered me the post.

I think I have said enough as to how very much I appreciated my time in Defence, but after six years, with my fourth Secretary of State, my sixth Chief of Defence staff, my third Permanent Secretary, I began to feel rather like the old man of the mountain. I also got a little tired of so many of the same problems coming round and round again. So I was very ready to make the transition and move from Defence to Energy on the 1st of October 1977. The previous summer had seen the particular celebrations of Jubilee Year where, especially, being with the Fleet at Spithead had been most memorable. I left my colleagues in Defence with a great tinge of regret, yet feeling I had done my bit as it were. Although the post in the Department of Energy only ranked as a Deputy Secretary, I retained my status, and also got the provision that to make my work in Energy effective I should stay there for three full years, at the end of which I would be eleven months over the age limit of 60. Three years as it turned out was a relatively short period in which to achieve much. The Chief Scientist in the Department of Energy has, in ordinary times, a curiously mixed set of responsibilities. One, which I found very interesting, but could not claim to be deeply expert in, was to advise my Secretary of State to approve or amend the research programmes of the great nationalized corporations, as they were then, in electricity, gas and coal. Another one was to keep some sort of liaison with the research effort of the privately owned oil industry. In particular, there was the task of encouraging them to buy and develop British engineering rather than overseas engineering for the rapid expansion of their effort in the North Sea. Yet another was to deal with the growing efforts in renewable energy research where one often had to liaise with and award contracts to the many quite small groups working in particular fields that had taken their interest. They were devoting their invariably considerable talents but sometimes very small resources to these tasks. This was all very interesting particularly perhaps in the field of wave energy. But the two jobs that really took my fancy were neither of them foreseen at the time when I joined the Ministry.

One came right out of the blue and was the International Fuel Cycle Evaluation (INFCE). President Carter had come into office early in 1977, very much concerned about what was then seen as the risk that the expected major spread of nuclear electricity generation throughout the world would encourage the proliferation of nuclear weapons to many more countries. While the aims of preventing such proliferation were very widely shared, the United States was wholly isolated in its choice of means. There had been an authoritative American report a few months earlier the conclusions of which were clearly embraced by President Carter. In this view the essential step was to prevent the separation of plutonium which was so necessary for the development of fast breeder reactors which themselves greatly improve the utilization of uranium, and so minimize the demand for it. It must be remembered that in those days there was great optimism about the spread of nuclear electricity generation. Therefore, many countries, not as rich in natural uranium resources as the United States, were deeply engaged in fast breeder research, notably ourselves, France and also Germany and Japan. All these countries had to import their uranium. American efforts in fast breeder research had not been as successful as those of others, including ourselves. And so what was certainly meant as very good anti-proliferation measure by the Americans was attacked as blatant commercialism and insensitivity by everybody else. Therefore this International Fuel Cycle Evaluation that was launched by President Carter had all the makings of marking a serious diplomatic split between the United States and her allies.

When INFCE was starting I was asked to co-ordinate and in effect lead the entire effort for the United Kingdom. Sitting in the Department of Energy but with my background in Defence, I was felt to be uniquely qualified for this task. Though I have never been a nuclear physicist, let alone a nuclear engineer, I was thought to be the right person for this task. I have never been one to shy away from responsibilities and so took this on. With innumerable meetings in many different places, notably Washington, Tokyo, London and Vienna, I had a very substantial task. I have always enjoyed diplomacy of which I had seen a fair amount in ESRO, and some in negotiating on weapons procurement for the Ministry of Defence. On the other hand my technical knowledge was really rather poor, and no doubt other countries would feel they had much better qualified people than myself. So I turned to Walter Marshall. I had displaced him in the Department of Energy and some resentment against me would have been natural, but Walter very readily agreed and put his immense knowledge and boundless energy into this task. He gave me total loyalty and a splendid level of support. I regarded the INFCE enterprise, which finished in 1980, as a great success since any split with the United States was avoided and it ended very harmoniously. I got on very well with the American Delegation, especially its leader, Gerard Smith and with Abe Chayes who played a most important part in all this. I was pretty active in the matter. As I was saying diplomacy suits me perhaps because I can articulate matters with great precision. This was appreciated so much that it was the wish of all the major countries that I should be the President of the final assembly at which the work could be completed. This did not in fact happen, because many of the other countries felt that the assembly should not be presided over by a person from a nuclear weapons state, and so a Japanese colleague became President. I had, in fact, worked very closely with the Japanese because we co-chaired one of the working groups, holding our first meeting in London and the second one in Tokyo, which I found a little awkward to get to. I claim that I then persuaded the Japanese that Vienna, which was the seat of the International Atomic Energy Authority, was just halfway between London and Tokyo and that we should have most of our meetings there. It was not quite fair but they regularly agreed except for one meeting which we held in Palo Alto in California. Vienna was, of course, a natural place, with the Authority there, with the curious consequence that I led the British Delegation not only in the negotiations, but through the streets of Vienna as well, which I knew from childhood. There were major policy decisions to be taken on what attitude the United Kingdom should take in the negotiations. I decided these in discussion with my colleagues in the team and, from time to time, reported to Ministers (Labour until early summer 1979, then Conservative). Invariably they wished me "More power to your elbow".

This task kept me pretty busy but long before it was finished, another job landed in my lap, perhaps more unexpectedly. There was great interest in renewable resources of energy. I should say that I detest the word alternative energies, because I think we will need all sources, of one kind and another. The energy problem will be far too serious just to chuck in some part as a replacement for another. There was special interest in using the high tides of the Bristol Channel to generate electricity, so I was appointed by Tony Benn to head the Severn Barrage Committee, to look at the feasibility, environmental consequences and economic benefits of a tidal barrage. This was a pretty formidable undertaking. It occupied me until the summer of 1981, and was very interesting, enjoyable and successful. It was a widely representative committee involving people from the Central Electricity Generating Board, the local authorities, ecologists, water authorities, parliamentarians. I am very proud that I was able to lead this large very heterogeneous committee to unanimity in the report we produced on this interesting issue.

In this work, as in my work on renewable energies, I had the help of Technology Support Unit, built up by Walter Marshall, and of Freddy Clark who acted more or less as my deputy, invariably full of energy, good sense and good humour. My work took me to coal mines, oil platforms, the landscape of Alaska, nuclear power station, facilities in the United Kingdom, the United States, the Soviet Union and France, to innumerable groups working on renewable energy sources etc! One subject I was especially keen on was to promote the gradual replacement of divers in the North Sea oilfields by remotely controlled submersible vehicles to avoid having to use people in conditions of stress and danger. Remotely operated devices seem to me to be very much the technology of the future. This interest also led to many journeys, as did the Severn Barrage.

I was anxious to get impressions on the spot on the relevant technologies. This worked out very well in the summer of 1979. I was going to Sydney anyway, to lecture at one of the science schools. My wife came with me on the first stage and we had a wonderful holiday walking in the mountains of the great National Parks of southern Utah, after which I went on to the State of Washington to observe the new French designed and made turbines being installed in the Columbia River, as these were the biggest turbines being designed and built for the low head, large volume that would be involved in the Severn Barrage. I then went on to Sydney and from there to Hong Kong, where I was taken to look at the High Island Dam built for a water reservoir, but facing the south east typhoon and so having to stand up to an attack by wind and waves even greater than would have to be endured by the Severn Barrage.

So I had plenty to keep me busy in my time in the Department of Energy. But three years was a somewhat short time to make real changes, particularly in the balancing of research and development effort between the Atomic Energy Authority and all the other tasks. The routing of advice was a little tortuous.

CHAPTER 14

Natural Environment Research Council

THEN ONE of those twists arose that led to another fascinating job. The Natural Environment Research Council (NERC) is one of the five Research Councils established in the mid-sixties and looks after a variety of fields of science, notably geology and the other earth sciences, oceanography, terrestrial ecology, hydrology, the British Antarctic Survey, and the like. Originally it had a curious constitution with a Chairman who was the senior officer but was expected to work only two days a week for the Council, and a Secretary under him who was full time, who was the Accounting Officer and, in many ways, the boss. It had been decided to change the system to a more normal one with a full time person as Chairman and Chief Executive. Of course, as the time approached various names were considered for this role and I, as one of the mandarins in science, was asked for my views on these possibilities. Eventually we all focused on one particular name, who we all thought would be extremely good for the purpose. We were just about to approach him when, it emerged quite suddenly, this person had become unavailable. There was great consternation as time was rather short, but at the end of the telephone conversation when the DES official in question talked about how awk-ward a problem it would now be to find somebody in so short a time, I said, "What about me? I will have to retire from my present job in a few months time, and though I am in a rather different field of science from those pursued by NERC, that kind of thing has never bothered me and I do not think it bothers other people. It is a task I might enjoy." This suggestion was taken up with alacrity by everybody concerned, and so when I finally retired from the Civil Service proper, on the 30th September 1980, I took up my post as Chairman and Chief Executive of NERC, with a four-year term of office at the end of which I would be almost sixty-five.

NERC counts, in the inelegant phraseology of the Public Service, as a fringe body, and so the age limit is not so tightly controlled, and in any case there was doubt whether it would apply to the top dog. Under the dispersal rules the Head Office of NERC, a distributed organization with many locations, had been moved from London to Swindon, though it kept a small office in London for liaison. For somebody who had been used to commute between Reigate and Paris, the trip to Swindon did not sound all that frightening. With my customary good luck, on the previously very slow railway line that ran from Reading through Guildford to Redhill, a fast service from Reading to Gatwick stopping at Redhill had just been instituted, and so daily commuting was entirely feasible. I could make good use of the many hours spent on trains, for reading papers for my work for NERC and, to some extent, for my own scientific papers. And so I found myself again as a Chief Executive with a new office, a new job and new friends to make.

It may perhaps be useful to discuss the United Kingdom's organization of science in general and the Research Council system in particular. There are basically four reasons why government, and therefore taxpayers' money, is involved in science. First there is the cultural argument for public support for science for its own sake without worrying about its utility. Curiosity-driven research is an adventure for mankind, and work in fields like astronomy or high energy physics (to mention two subjects close to my own heart) has indeed influenced the public's attitude to the place of man in the universe. This cultural claim on the public resources is similar in kind and in magnitude to the claims on behalf of museums, art galleries, the National Theatre, major orchestras etc. In this field, while parliament naturally has to decide how much money should be spent, scientists themselves are the only people capable of choosing in detail what the money should be spent on. Second comes support for fields of science that seem likely to become economically useful one day, but where this lies too far in the future for the eventual user to fund the bill today. Where this time horizon lies differs from industry to industry. Pharmaceutical firms are prepared to invest in very long range research, farmers basically leave the funding of research to others. If government does not pick up the bill, the sector concerned and perhaps the economy as a whole will suffer. Here government usually depends on the industry concerned to give advice on what may become valuable for it. Often its interest will be shown by shouldering a portion of the cost. Third come the fields where the government needs science in order to discharge its own functions. For example, the government needs to have an under-standing of the movement and recharge of ground water to be able to regulate waste disposal, water extraction and water quality. Similarly government needs to know about organisms in food that may cause diseases in order to institute regulations that ensure an acceptable standard of safety. Both governmental needs and scientific possibilities need to be judged in order to spend public funds in this field to good effect. Fourthly there is the educational aspect of science. Not only is a modicum of scientific knowledge and of how science advances necessary for responsible citizenship, as should be ensured by the country's schools, but at the highest levels of responsibility there should be people with enough understanding of, and sympathy for science to enable them to contribute to good decision taking. Add to this the need to educate the research scientists of the future, and the task of government is seen to be very major. A sizeable portion of this task is shouldered in the United Kingdom by the Research Councils, semi-independent authorities whose members are appointed by the Secretary of State for Education and Science. They are at one remove from his department (DES) yet basically belong to it. The first task of a Research Council is to serve university research. This is done by the support of students for higher (i.e. post bachelor) degrees, by the provision of funding for research projects, enabling the researcher to hire assistants and to purchase equipment, and by providing major equipment (e.g. research vessels or telescopes) for the use of university researchers. Secondly the Councils own and fund institutes and units to carry out work that by its interdisciplinary nature, its need for continuity, or its technical demands, is deemed not to fit into the university system. Almost every country has its own geological survey, its own centre for epidemiology, its own meteorological office etc. Other institutes, such as astronomical observatories, belong to, and are maintained by, a Research Council. They have there resident Research Council staff who maintain the instruments, improve them, and design, purchase and install new ones. Naturally the astronomers among the resident staff pursue their own researches there, but the observatory exists largely for university researchers, senior or junior, to carry out their programmes.

All the relevant fields are divided up amongst five Research Councils. I had served on one, the Science (now called Science and Engineering) Council which covered my own fields of work in theoretical physics and astronomy, and now found myself in charge of a very different one. Yet there are great similarities between all forms of scientific research with its frustrations, its need for persistence and patience, its elations, its dependence on asking awkward but revealing questions. So I was received with the greatest friendliness and interest by all concerned. NERC responsibilities were essentially the fields of science in which the primary evidence has to be gathered out of doors: geology, ecology, marine and freshwater biology, hydrology, all the fields of science that can only be studied in Antarctica (through the British Antarctic Survey, a component institute of NERC). Logically all the atmosphere sciences including the Meteorological Office should have come under NERC, but in 1965 the Defence interest in weather forecasts had left the Meteorological Office with the Ministry of Defence. This led to some awkward boundaries which however could be managed reasonably well. I threw myself with the greatest interest into learning to appreciate the needs, the problems, the aspirations and the resource requirements of these sciences. Learning from visits to active laboratories and field sites I found most stimulating, and indeed during my time with NERC I published a paper on ecology.

NERC had serious problems arising from an earlier series of decisions with which I had had some connection. When following the Trend Report the Research Councils had been set up in the mid-sixties, their responsibility was for subject areas being pursued in universities, and in national and international institutes. In all Research Councils a significant part of the work was to support other activities of the Government. It was left to the Council itself to formulate what were considered to be important research areas in the national interest in, for example, agriculture. In the early 1970s, Lord Rothschild, working from the Cabinet Office, produced a Report suggesting that there was something fundamentally wrong with this. Members of Research Councils, however eminent, were not the people to take political decisions as to which kind of research was required to support Government policy. Therefore, a certain amount of the funds of Research Councils should be handed to the relevant Government Departments, who had to equip themselves with Chief Scientists and their staffs in order effectively to commission work from the Research Councils.

This proposal, to make the Government Departments themselves choose the research needed to fulfil their functions and provide the required funds, seemed to me entirely logical and I supported Victor Rothschild, whereas the bulk of the scientific community hated the whole idea, not trusting as it were the competence of Government Departments. Victor Rothschild's idea was that Government Departments had to be equipped with the required scientific expertise. The chief stumbling block to the reform was the Medical Research Council, which not only had been outstandingly successful in supporting superb researches in many fields, immunology, molecular biology, and so on, but where also the Ministry of Health was not particularly keen to take an interest in these matters. Though these funds were in fact transferred to the Ministry, before very many years were out, this was abandoned and the Medical Research Council again had full control of all its fundings.

NERC was involved with quite a large number of different Government departments: The Department of Energy, because of the underlying geology that was needed for oil exploration and production in the North Sea, for supplementing the work of the Coal Board's own geologists, and so on. The Department of the Environment, in its planning work for the ecological results, of work about the movement of ground water for water companies, about the stability of roads against land slide, about coast protection and so on. The Ministry of Agriculture, Food and Fisheries was, of course, much involved in matters of oceanography as regards the fisheries, and naturally there were also considerable links with the Scottish and Welsh offices, as well as with Overseas Development.

After the Rothschild reforms at first things went well in that as government departments became more scientific, the demands they made for research and were prepared to pay for increased sharply throughout the seventies leading to a great growth, in particular in the staffing of the British Geological Service. But with the constraints on public expenditure in the eighties these demands fell very rapidly, particularly as regards the Department of the Environment. I am proud to say that during my time as Chief Scientist of the Department of Energy an arrangement between the Department and NERC was negotiated that was very advantageous to both sides but elsewhere there were much greater troubles. The boundary line between applied research and pure research has always been fuzzy and if departments took too narrow a view of what they were prepared to fund then the background for the work they demanded simply was not available. Moreover NERC suffered from having appointed people on Civil Service conditions with tenure, with money for their salaries depending on short term contracts. When the contracts shrank there was not enough money left to pay these people or more correctly it had to be borne by the already hard pressed science budget of NERC, which had to go on the salaries of people who were no longer required in such numbers instead of on equipment. But on the equipment side too the demands on NERC expanded very rapidly, for several of these "outdoor" sciences that in earlier days only needed very simple tools now demanded highly complex equipment to examine rocks, to examine plants and the like. The areas of oceanography with the need for research vessels and of the British Antarctic Survey were by their nature expensive. In my view Rothschild's principles were basically right. Where mistakes were made was not in the principles but in their application. To have hired people in large numbers on the basis of soft contracts is always known to be a bad system, and NERC and the Geological Survey in particular had suffered greatly from this. Where I felt and feel critical of the Rothschild proposals is that they made no provision for any transfer of money from the Science Research Council, since renamed the Science and Engineering Research Council (SERC). This meant that it was left to the SERC itself to decide how much money to take from the many branches of science for which it is responsible, such as physics, astronomy, chemistry, biochemistry, and so on, and transfer this to the engineering side in view of the undoubted need for this country to modernise its engineering industry. In my view, a modest amount of money, corresponding to engineering applications, should have been transferred in the early seventies from the Science Research Council to the Department of Trade and Industry, and then the required expansion of engineering work would have come from that side.

Another department with which we had close connections was Overseas Development. In a number of cases, aid was given by supplying some of our staff as experts to support the work of the developing countries' own scientists or civil engineers to work there. Also, when funds came to a civil engineering consultant to design a dam in some part of the world, they were quite likely to sub-contract to us for some geologists and hydrologists to tell them about the stability of the possible dam sites, about the water regime in the river, and so on. I was particularly intrigued when I saw the system at work in countries that were quite strong in these fields themselves, but could do with a little extra help. For example in Indonesia, the Geological Directorates of Indonesia were given help by our sending out people there to assist their staff, to bring other experience in, and generally speed up an ongoing enterprise of mapping the major natural resources of Indonesia. Similarly our hydrologists were often employed in arid areas of Africa and elsewhere to locate aquifers from which clean water might be obtained, such a necessity in so many parts of the developing world. Other groups went under ODA to Bolivia and Ecuador to do geological mapping of the higher Andes. All this was very exciting work for the people concerned.

In spite of these interesting outlets, some parts of the organization were distinctly depressed. The fall in commissioned income put great pressure on manpower, and people feared they would be declared redundant. Moreover, in order to avoid this, they were under-equipped and very often we had the people, but could not let them work, for this would have added to the costs. So we had a difficult time.

I saw it as my task, not only to do the most vigorous propaganda for NERC wherever I could, on the Advisory Board for the Research Councils which apportioned the Science vote and elsewhere, to try to attract contracts from home and overseas; to strengthen collaboration with civil engineering consultants, and others. Getting contract income from the private sector was important. Though it could not lead to a major financial breakthrough, yet it could be immensely useful. It often required a change of attitude which again I saw as part of my job to encourage. Above all I felt that NERC staff and University staff and students supported by us should see me. I did not, as the Chief Executive, want to be seen as a remote paper pusher. So I travelled far and wide, visiting our own people at their institutes, in their field work, and visiting many universities where we supported work. I think in my four years I visited every one of the forty-three sites where NERC staff were working in the United Kingdom, and several abroad. I must have visited twenty-five or so University groups and generally was pretty active. The one thing that I missed out was to go to Antarctica. I would have loved a brief visit, but that was not really possible; our bases could not take long distance aircraft and so I would have had to go on the supply ships. Not only am I bad sailor, but logistic considerations dictated the moves of our supply ships, and they could not make special dashes to please me. So a trip of less than about six to eight weeks would not have been feasible. Not only did I doubt whether I had the patience for this, but I did feel that my work at home could not permit me such an excursion.

With all the meetings, visits and discussions, I had a little time for other activities thanks to the immense capacity for work of my second in command, John Bowman, the Secretary to the Council. He became a friend and left NERC only very recently to be the first Chief Executive of the National Rivers Authority. In addition to writing the odd scientific paper, I served in 1982 as President of the Association for Science Education for whose work I have the greatest admiration. I accompanied the then Secretary of State for Education and Science, Sir Keith Joseph, on a most impressive visit to its headquarters.

A little earlier I became (and still am) President of the Society for Research into Higher Education. Though I am much interested in its very much needed work, I have not been very active in this post.

One of my strong principles has always been, that consideration for the future must come first. A year or two before I joined NERC, the organization's Council had decided that its finances were in too low a state to allow it to renew any of its research vessels, on which, of course, our work in oceanography so largely depended. This was not to my taste. Though we were even poorer in my time than when Council had taken this decision, a year and a half after I joined NERC we had signed a contract for a new research vessel and she was commissioned shortly before I left the organization. I am intensely proud of the fact that in Charles Darwin the organization has a first-class new research vessel, albeit originally somewhat under-equipped. That will no doubt be mended in time but we no longer have to rely on ships that will be phased out in a relatively short period of time.

Through all my years in the public service my work was made effective and enjoyable through the efforts of the private office I invariably had. The service appreciates that to make its top officials, just like ministers, use their time and energy to good purpose, a private office of several people is required. I have been served with splendid loyalty and industry by all who worked for me. The head of my private office was usually a young high flyer for whom this was an excellent and most instructive step in the career, lasting often from one and a half to two and a half years. I have had the pleasure of having had a succession of outstanding people in this position, people who could think themselves into my outlook and helped me enormously. I am most grateful to them all.

CHAPTER 15
Churchill College

IN SPRING 1982 I got a letter from Churchill College, Cambridge, asking whether I would be interested in my name being put forward for its mastership. This was by no means the first such invitation I had had, but I had rejected all the others readily. First, I thoroughly enjoyed having a full-time, demanding, managerial task, whereas the headship of a College is certainly not full-time. Secondly, great though the honour is, my memories of the overall administrative atmosphere of Cambridge were not over happy. Third, we were rather committed to the maintained school system, and Cambridge was to a large extent populated by products of independent schools. Yet the invitation from Churchill looked different to my wife and myself. First the timing: my term of office at NERC would end in September 1984, just before my sixty-fifth birthday. Would anybody wish to employ me after this in a major managerial task? It was already a little embarrassing for me, in my early sixties, to deal with the age limit of NERC which, for ordinary staff in the main was sixty. And in an organization that had to shrink I could hardly ever permit anybody to go on beyond this, even in the cases where I was really keen to do this. For me in my early sixties, to tell people that at sixty they were too old to serve, was a little embarrassing. This embarrassment would have become severe, had my appointment been prolonged into my late sixties. Next, the newness of Churchill was an attraction. We felt that the barnacles that have grown up on the older ships would not be so strong there. Yet when we both fixed the date for visiting Churchill we were by no means sure that this is what we wanted to do. Since we had worked together on the constitution of the stars in the late 1940s, Christine and I had had very different careers. After the family she had gone in for teaching, and later became a Magistrate on the Reigate Bench. Though, of course, I would be the Master, yet the spouse of the Head of a College should ideally take an active part in the matter. So there was a prospect of us both being involved, and she, also, was approaching normal retiring age for a teacher, though not for a magistrate. Being at Churchill would give her much opportunity for continual contact with young people.

Our visit to Churchill on the early May Bank Holiday of 1982 totally charmed us and won us over. Churchill is a college modern in outlook as well as structure, a college full of friendliness, a college where the large majority of the undergraduates come from the maintained sector, a college with strong international links. What more could we ask for? Of course, there was a retirement age too, namely the summer after the seventieth birthday which, in my case, would be July 1990. But this was far beyond what I could possibly hope for in the public service. The appointment is a Royal appointment, and the College can do no more than suggest names. Yet we left on such friendly terms that we thought there was quite a chance that the College would wish to put forward my name. Much more speedily than expected I had a letter from the Prime Minister asking me to take on the post. There were still a few rough edges to be dealt with, notably the overlap. The Mastership of Churchill became vacant through the retirement of my predecessor, Sir William Hawthorne, in the summer of 1983. Will and I had known each other for many years, both in Defence and in Energy, so there was the additional pleasure of succeeding somebody whom one knew. But my term of office in NERC ended only in September 1984, nor was Christine keen to give up her teaching too soon. So it was agreed that for the first year of my Mastership we would be non-resident. As I like to put it, I was an absentee landlord. The College readily agreed to this although it threw a tremendous burden on my delightful and very able Vice-Master, Dr. Jack Miller. And so, long before my NERC job ended, I was already fixed up for the next six or seven years. During the year 1983/84 indeed we both had our heavy jobs, I in NERC, and my wife in the Reigate Sixth Form College, and as a Magistrate, but we spent most term time weekends at Churchill.

I should mention here another aspect, where Churchill as a College and we as a family (as will be related later) have followed the same route, and that is in encouraging young people to do a year's work between school and university. Churchill College insists on such a year's work for all its engineering students, who form nearly a quarter of the total, and encourages it for the others, though there is not the administrative strength to organize it outside engineering. But perhaps another quarter of our intake arrange it for themselves.

When we finally moved to Cambridge in September 1984, that was just thirty years, almost to the day, since we had left. Much as I had enjoyed my years away from University life (effectively since 1967), to be again in a place full of young people was a delight, and we have been having a very happy period at Churchill College.

To have become Master of Churchill College was a particular pleasure for me, with my historical interests. To be the Head of the national memorial to such a great man who was so influential in my time, and who I admired so greatly, was wonderfully to my taste. When I speak to our young people, at our Annual Founder's Day, I am only too conscious of my responsibility as being the last Master of Churchill College who experienced Winston Churchill's leadership as an adult. We have also greatly enjoyed our contact with Churchill's family, notably with his younger daughter, Mary Soames (whose much liked and admired husband Christopher unhappily died during our time at the College), and his grandson Winston Churchill M.P.

Perhaps it would be useful to describe here briefly the significance of Cambridge colleges and the position of their heads. The Colleges are private institutions of charitable status, effectively educational trusts controlled by their trustees (the Master and Fellows). The tasks of Cambridge are divided between the University (which is grant aided by public funds) and the colleges. The University of Cambridge is the body responsible for giving degrees, holding the necessary examinations, providing teaching through lectures and is responsible for laboratories for teaching and research. The colleges are responsible for selecting their students from the many who apply, for teaching in tutorials (supervision) of rarely more than two students at a time, for advising them on courses of study and of their progress, and are responsible for housing them and for providing meals. Moreover, the colleges are the focus of the social and sporting life of students (particularly of undergraduates). The students pay the colleges fees for the academic help and advice, and dues for their board and lodging. Undergraduate college and university fees are fully paid by local authorities for all students resident in the United Kingdom, and in addition there is a grant for board and lodging dues, but only if the parental income is modest. Postgraduate students from the United Kingdom get a grant from the relevant Research Council covering both fees and living costs. Thus though a college does not receive a grant from public funds, a good part of its income is payment from such funds in respect of services rendered to students. However, colleges have also the income from their endowments which are very different in amount and importance for different colleges. The Fellows of a College are elected under various titles, but the two most important ones are because of the work they are doing for the college (in teaching or administration) or for eminence in research. A particularly significant group of Fellows are the Junior Research Fellows, normally elected in their twenties for their early researches. This is a very important step for aspiring young academics immediately giving them recognition and status. It will be recalled how elated I was at my election by Trinity in 1943.

A college is a very important social unit for its Fellows who meet over dinner or who may live in College where some of them do much of their teaching and administration. However, many of the Fellows are also University teaching officers, and most University officers are also Fellows of Colleges. University facilities (laboratories, buildings) are usually the centre of research activities.

The Master of the College is its titular head, and chairs all its important committees. He is in no sense a Chief Executive, the College being a very democratic unit. He represents the College externally and also has important social obligations for the communal life of the College, at the senior as well as at the junior level. In these the Master's spouse can play a very major role.

What is particularly enjoyable at Churchill College is its very international character. Amongst our Fellowship of a little over a hundred, there are usually ten or so from overseas, and several other senior visitors from other countries. About half of our 190 or so postgraduate students come from overseas, but almost all our 380 or so undergraduates are from this country. Part of our duties is to make overseas visitors feel at home, and indeed most of them come and visit the College whenever they come back to this country, even if it is years since they resided here.

The Master also tries to add to the wealth of the College by soliciting donations. Though Churchill is fortunately placed with quite a reasonable endowment and buildings that allow us to house all undergraduates on our site, we cannot do the same for all our postgraduates. In my last year here we received a munificent donation from the Danish Maérsk Foundation that will enable us to offer accommodation to virtually all those postgraduates who would like it. Moreover, the Maérsk building will allow us to operate year round (and not only out of term) as a study centre in the increasingly important area of postexperience education.

But the most significant thing a College can offer is life in a civilised academic atmosphere through which we can offer to the ablest minds so agreeable an atmosphere that they acquire a permanent liking for the institution. It is a great privilege and honour that we have been able to contribute to this.

As I am writing this, my retirement is barely a year away, but there are many other things that keep me busy. Shortly before I came to Churchill I accepted the Chairmanship of the International Federation of Institutes of Advanced Study, an organization whose title is perhaps a little more highbrow sounding than it is, but whose task is to organize research projects, involving both the natural and social sciences, and both industrialized and developing countries. This is very much to my taste, and involves me in a good deal of travel. In Cambridge the College kept me pretty busy. For much of my time I was also Chairman of the Colleges' Committee, which brings all the Heads of Colleges together, became Chairman of the Faculty Board of Education, did a bit of lecturing and so on.

During this period I also served for two years as President of the Hydrographic Society. I enjoyed the invitation to do so especially because I am so ignorant of the subject. However, I believe I was quite useful to the Society through my acquaintance with observations from space. In any case I thoroughly enjoyed my contacts with the hydrographers.

In the College there were and are some knotty problems to deal with, but I think we are facing them all in a spirit of good will and confidence in the future.

Throughout all this period I have never neglected my scientific work. Being a theoretical person, all I need for it is pencil and paper. Even after I retire from the Mastership I expect this will keep me pretty busy. In my most energetic managerial days, the only time I had for my own research was effectively on aeroplanes and on trains. But I did manage a tolerable output of papers in all those years. One of my anecdotes about this period is of being stuck in an European airport, through the incompetence of an airline, for most of a day. When I published the paper resulting from my work there, I wondered whether to dedicate it to the airline in question!

The considerable duties at the College, to bring people in from the outside, to make our many overseas visitors feel welcome, to encourage our students, are all most agreeable. Yet I think after seven years as a Master, six of them in residence, the College will have a right to see a new face, and I think I will probably have given as much to it as I can. So I am continuing to look forward to the future with confidence and in expectation of agreeable things to do.

CHAPTER 16

Reflections

FROM QUITE early on in life, as I have mentioned, I was full of confidence in my ability and productivity in applied mathematics. I think it is this self-confidence that made me a good teacher, because self-confidence is so essential to a good relationship between the teacher and the taught. Yet, to my surprise, in my career I have done much more than research and teaching in my particular field. I have been asked to hold all sorts of posts, and people generally have made very nice noises about how I performed in them. What have I got that gave me so much wider a life and experience than most of my academic colleagues? Clearly there are some basic personal traits, such as my sympathy for and with people, my wide interests, my enjoyment of communication. But I think there are some other factors too. I have not, for a long time, over-estimated my intelligence. It is not just that my children, for so many years, have reminded me of all my stupidities. It is I myself who very often realize that I should have seen a point and did not. Now this has had two consequences. One is that it has made me rather honest, not necessarily from purely moral grounds, but because I do not judge myself as anywhere near clever enough to carry through an untruth successfully. In my relations with other people, in management, what people above all want is that they can trust you, and I think I have been able to inspire some trust. But another aspect is that I have never been afraid to ask questions though they expose my ignorance outside my own narrow domain. This may be due in part to my wide interests in so many fields. It may also be connected with my experiences during wartime, when I, the most theoretical of people, found myself involved in the engineering of naval radars and simply had to become more practical. In addition there was the influence of Tommy Gold who taught me to regard intelligence as non-specific. All this attracted me to wide areas and to try anything that came my way. It is fortuitous that these activities have always been in the public sector. I would have been interested to work in the private sector, too, but somehow none of the potential openings ever materialized.

Of course my own craft of Applied Mathematics is essentially a tool, a tool that can be very widely applied, as I certainly have done. I have published, not only in the standard scientific journals for my field, like the *Monthly Notices of the Royal Astronomical Society,* or *The Proceedings of the Royal Society A,* but in engineering journals, and even in biology, where, not many years ago, I published a paper on mathematical ecology. In the early seventies I tried to encourage people, through a Government Committee I was asked to chair, to move between university, Civil Service and industry. It turned out to be arduous to organize this, but the ones who made such changes without exception enjoyed them and profited by them. The depressing thing about this effort was to learn how our society seems to work in the opposite direction and tries to confine people, through their trade union, through their professional association, through their habits, to stick in a rut. A rut, as I like to say, is just like a grave, only longer. When I had some enquiries carried out about the experiences of people in the course of this Committee work, many said "The only interesting thing that ever happened to me was during the war", which is a pretty sad commentary on our society. For what did the war do for these people? First, certainly, it made them change from what they were doing, and secondly, it gave high responsibility to very young people, something our society rarely does in peacetime.

In my visits round industrial companies, it was easy to pick out the ones that worked well because they gave responsibility to younger people, and those that failed to do so. (A little story about my visits to small industrial research laboratories: where on more than one occasion the Director greeted me shamefacedly and said, "Very happy to see you come sir, but of course what we do here isn't really science, it's only trial and error." To which my unkind reply was, "Oh, I didn't know there was anything to science other than trial and error.") Change invigorates one and is a major element in keeping one effective. Indeed I would go further. The outstanding human characteristic, the one where we differ most from other animals, is our adaptability. We alone can populate the wet tropics and the deserts, the Arctic and the temperate zones. We alone can be vegetarians or meat eaters, or live on the products of the sea. And so much in our society tries to prevent us from exercising our human genius for adaptation and change.

I am more optimistic for the future when, with lifelong learning and mature entrants to higher education, there is more chance of people not sticking to their last, and that, I think, is what is half the fun of life.

Quite apart from my attitude to change, my enjoyable career owes much to my good health. While I was not particularly sporty as a young man, and have had my colds and influenza and the like, and occasionally even back trouble, it was later in life that I suddenly began to realize that I really was remarkably fit and untiring. In particular, travel was something that never wore me out as much as other people. On a job I could do several hours work in one place, fly to a second one, work there, go to a third one, and work there again. Only recently, just before I was sixty-nine, I woke up early one morning in Toronto, took the breakfast flight to New York, saw my sister, took part in the Board Meeting of the Winston Churchill Foundation of the United States, went in pouring rain to the airport, flew home, arrived in the Master's Lodge a little after ten o'clock, and within half an hour was chairing the meeting of the Heads of Colleges. That afternoon we had a party of students for tea, and that evening I hosted a Reception and Dinner for new Fellows.

It has been like that for quite a while, and it began to make me, perhaps from my early fifties, childishly proud of my good health. It was not just that I could go skiing, it was that I could be untiring. Of course that had a lot to do with my job. One feels tired when things are boring, but whenever you do interesting things, whenever you are put on the spot, you wake up immediately. Adrenalin is a habit-forming drug and makes one very fit. Yet my childish pride in my health was spotted by various doctors. In the Senior Civil Service, every few years we had a medical examination. When one of the doctors finished explaining to me all the excellent features that he had discovered in my tests, he said "I am afraid the regulations don't allow me to issue certificates of immortality." Another, when again he had been going through how excellent every test had been, and saw me preening myself said, "Of course, there is no merit in this." But I am sure I have been greatly helped by being so particularly fit, especially in middle age and since. I suspect it was all helped by my dislike for taking medicines. I have never used sedatives or any regular medication, and am irritated by having to take anti-malarial pills when I visit the tropics. However when my back is troublesome I find a combination of aspirin and red wine very helpful.

I was also very lucky by never having become a smoker. When I was an undergraduate, in the late thirties, most students smoked. The effect on health was not then known, and the cost was modest. Though I tried once or twice a cigarette, a cigar, or even a pipe (which I thought would suit me) I hated every second of these exercises. Not being very persistent, I stopped them. The only way I ever enjoyed tobacco was to take a little snuff after port in my days as a young Fellow of Trinity, and even this was not all that attractive to me. No doubt my never smoking has been a contribution to my good health.

While my energy and drive have frequently been commented on by others, I have equally admired and enjoyed very energetic people. I have already mentioned Harry Messel, for many years Head of the School of Physics at the University of Sydney, who is a friend of long standing and often asked me to come there to lecture at the Science Schools he arranged. On my first visit to Australia, a very full and demanding lecture tour for the British Council, the only occasion when I felt worn out was on the day that had been set aside for me to relax with Harry!

Through him I met Robert Maxwell, whose energy is as legendary as his war record. I had first heard of him in the days when I was Secretary of the Royal Astronomical Society. He then became known as able to publish scientific books of great importance, but of possibly modest size of potential readership. This very significant contribution to science naturally made me admire him. After we met, we co-operated happily on a number of scientific and educational ventures. Christine and I went to his sixtieth birthday party at his home, Headington Hill Hall in Oxford. It was a splendid and memorable occasion.

From my early days, influenced by my mother no doubt, I had a distinct dislike for religion. As the years rolled by this became both stronger and more tolerant. I have been active in the Humanist Movement at least since the early 1950s, and am now the President of the British Humanist Association. I have written about these matters in various places, but let me just say here that Christine and I have never felt a need for religion. On the contrary we have been strong in our view that religion tends to be divisive, and that it is not the task of the Humanist to quarrel with anybody's personal religion, but to try to bring secular considerations to the fore, since these bring people together while religion and, particularly, a belief in the absolute truth of a revelation is such a divisive matter.

As a matter of course we got married in a Registry Office in Cambridge in 1947, and all our married children did the same, and nothing else ever occurred to them. Our youngest indeed got married in the same Registry Office where we had got married almost forty years earlier. Humanism takes a significant amount of time for us. Christine is very active, particularly on the Humanist Education Committee; I go to speak to Humanist groups in many parts of the country. We always go to the annual Conference and to various other events, notably of the local Humanist group. The Reigate one Christine helped to found and we were both very active members of it. We support the Humanist group in Cambridge.

A relevant anecdote is of the mid-fifties, when we had a conference in Oxford to consider the problems of non-believers in different countries. These were sometimes quite surprising, such that even in a country as broadminded as Sweden, any Swede born there is automatically a member of the Church of Sweden. Then a young man got up and said he came from the middle of Ireland, where the Church of Rome had a tight grip on education all the way up. It was very strong and thoroughly indoctrinated everybody. He had described it, of course, from personal knowledge, and had us all practically in tears until somebody called to him, asking how growing up like that he became a non-believer. Without a moment's hesitation the young man said, "That was a miracle."

Since its earliest days I have been involved with what is now called the International Science Policy Foundation (originally the topic was called

"Science of Science", and "International" was only added a few years ago). Its interest in the borderland between science and government and, to some extent, the sociology and philosophy of science, has always intrigued me. When it all started I had had no inkling that I was to hold such major positions in the public service. Inspired, driven and organized by its Director, Maurice Goldsmith, it has arranged numerous meetings which explored such issues and inevitably made the participants take a wider and more informed view than they had had when they came. In this way and through its publications it has been quite influential.

It is perhaps worth while here to expand somewhat on the evolution of my political attitudes. It was during my time at NERC that I first joined a political party. I have already mentioned the relief with which I greeted Churchill's accession to power in 1940, not only because of the assurance it gave as regards the prosecution of the war, but because it truly was a revolution, consigning to history the attitudes of the pre-war "National Government". In 1945 I was greatly heartened by the success of Labour. This was not out of disrespect for Churchill, but because he and his close associates, like Eden and Macmillan, seemed to have become prisoners of the unreconstructed rump of the Conservative party. Much of what Labour did and achieved under Attlee (a man I admired greatly) I liked, but I was greatly disillusioned when Sir Stafford Cripps, newly made Chancellor of the Exchequer, tried to correct the persistent nagging imbalance of trade by ordering industry to reduce its capital investment. This disastrous decision (so much at variance with Cripps' widely purveyed ascetic image), while for the moment preserving the standard of living, sold out the future. I ascribe much of Britain's troubles in the subsequent thirty years to this choice of preferring a modicum of comfort for the present to building for the future. Contrasting attitudes to the near-term versus the long-term has throughout seemed to me the most legitimate division in politics. The main advantage of the central planning advocated by Labour is surely that it *can* be used to put more resources into building for the future, and the failure to do so was a loss of nerve at a vital moment. Of course I recognize that in 1947 the standard of living was very low, and it would have required a great act of confidence in the people of this country for a government to have lowered it deliberately to ensure investment for the future. This is just what, it seemed to me, a Labour government had been elected to do, and what it failed to deliver. The opposite choices were made in West Germany, and we all know the result. Yet many other achievements of the Labour government were much to my taste, notably India, the National Health Service, town and country planning etc.

After my naturalization in 1947 I had the vote. Though I often supported Labour, my enthusiasm was very limited. But I appreciated early on the enormous difficulties of the choices politicians have to make.

Especially after arguing, as related, on behalf of King's College, London, my respect for the politicians of all persuasions has been great. This stood me in good stead when I entered the public service. I worked very happily with all Ministers I served, Labour and Conservative. The personality of the office holder seemed to me to matter far more than the party. Leadership, power of decision, managerial competence, ability to take the main issues on board, appreciation of institutional considerations and prejudices, these are of the essence in effective government. Of course as a civil servant one has to understand that ministers of different persuasions are under different pressures and have different difficulties in explaining their decisions to their followers. But once one has understood this, working at ministerial level is no different, though perhaps more satisfying, than work at any other level. The media, with their stress on great gladiatorial contests, in politics as elsewhere, manage to hide how much of the work of ministers is simply good management and sound common sense rather than ideology, and this goes at least as much for the ordinary parliamentarian. Nor do the media make it clear how fortunate this country is in getting such excellent people, with an overpowering sense of public duty, to enter political life. I have previously stressed my admiration and liking for the Secretaries of State I served, but I could add many more names to the list: Ministers of State, Parliamentary Under Secretaries, private M.P.s, and my many friends and former colleagues in the House of Lords. Yet when the Labour Party, around 1980, slipped to the left in a manner wholly unacceptable to someone with my views on defence, when ideology seemed to become unduly important amongst Conservatives, the beginning of the Social Democratic Party aroused my enthusiasm. For the first time in my life I joined a political party, as did my wife. She, indeed became much more involved than I, and fought two County Council and one City Council election as an SDP candidate. Though she never won a seat, on one occasion she came very close to it. This was sad for, particularly on the County Council, her experience of education and as a magistrate would have made her an influential and beneficial member.

The ill-advised merger mania of 1987 was a disaster for the middle ground in British politics and ended my enthusiasm though I am still a member of David Owen's SDP.

But to come back to my own career, a feature that I did not expect in myself and only discovered when I was thrust into chairing meetings and into general administrative work was that I enjoyed it. No doubt some of this was due to my being, to my surprise, rather good at it, some to the sheer pleasure human contact always gives me.

In ESRO I first began to learn what management was like, and what was expected of me in this direction. I had not appreciated until then how essential communication was, and how ill-equipped most of us are in this field. Well before my time in ESRO I had already been worried by the extent the education of a scientist fails to include training in communication skills, and had given my Grenada Lecture on this topic in the 1950s. After all in science, nothing is done until it is published, nothing has happened until other scientists take note of what one does, usually through personal contact as much as through the printed word. Yet there are still people walking round who claim that science deals with facts and not with people, perhaps giving this as a reason why they do not follow an education in science. What nonsense! Without communication, without co-operation, without the criticism of others there would be no science.

Karl Popper, of whom I am a great admirer, has described science so well and the human contact that is so vital to it. Yet we all have difficulties explaining ourselves. When a student writes his Ph.D. thesis, he is perhaps doing the first long, coherent bit of writing since he was 16, and usually also suffers from overestimating the rate at which his elders can understand new work. I might mention here that when I was Secretary of the Royal Astronomical Society and a young person was due to give a paper, my advice was: "You are addressing some very distinguished astronomers. Please speak to them as you would to twelve-year-old children." Only the few who followed my advice stimulated a good discussion.

But to come back to management, almost every problem that an organization suffers from arises from insufficient communication. For a compulsive talker like me, this was a relief to realize. I appreciated how much I could do, simply by going round talking to people and listening to them. Almost every failure in management is a failure in communication. Of course, decisions have to be taken, but decisions arrived at after ample communication are not only better decisions, they are much more widely accepted and understood. And so the succession of managerial posts I had suited me well. I like to meet people, I like to communicate with them, and I enjoy it when I see a disparate group of people getting to grips with a problem from very different points of view, acquiring a great deal of cohesion and mutual respect and liking in the course of it. In every post I have had, I like to travel, to see the people on the spot, to get the atmosphere of how they worked and what their problems were, to enjoy their particular branch of science, and to question them on it. I often make suggestions, which are not always bad and which very occasionally are found quite helpful.

I have not talked about our family in any way commensurate with their importance to my enjoyment of life. To have had five children is in itself a witness to the pleasure they have given us. One of the advantages of a large family is that one can observe and admire the way the genetics of bisexual reproduction produces real individuals, unlike anybody who has ever existed before, and see the differences (and the likenesses) between children of the same parents growing up in the same home. The undefeatable logic of small children has always given me great pleasure, as has Lewis Carroll. For example, I recall one of ours, very keen on the road, asking "Why is it that we so often see a sign 'road narrows' and hardly ever one 'road widens'?" I am sure that what one tells one's children to do has little effect, but they can never shake off the influence of the way their parents live. None of ours knows how to do a job other than to the best of their ability, none of ours can live without a good pile of maps

As I have already said, they all went to the local state schools. While the eldest went straight from school to university, (she now lives in the North Pennines and works on countryside management and interpretation), none of the others did so. With the second, Jonathan, it all happened in quite an amusing way. He was ready to go to university, while I was still with ESRO, commuting to Paris. He was keen to spend a year far away from home, having a job in India or in Brazil. Fine we said, that will be very good. The summer months dragged on without anything happening. In September, on a Thursday evening, I rang home from Paris, as I often did, and Jonathan wanted to talk to me. "Look", he said, "I have decided I cannot get myself a job far away. You know so many people on the Continent, could you find one there for me?" "Yes, of course," I said, "I will try." "But," said Jonathan, "I really must tell the University whether or not I'm coming this year or next. Could you have it settled by the time you come home tomorrow night?" Now this was a somewhat tall order, particularly as at this stage the United Kingdom was not yet a member of the European Community, so it was not only a job that was wanted but a resident's permit and a labour permit. Enquiries from most of my colleagues next day elicited the response that they were sure the whole thing could be arranged, but it would take at least six months to do so, which of course rendered it all useless. But my Italian colleague, Umberto Montalenti, said that he had a friend who ran a software house in Milan and with it a training site for young people, and because of the training aspects it did not fall under the regulations. He then rang his friend in Milan, who he reported was most interested in my son. I was a little suspicious. I feared that the knowledge of an English eighteen-year-old in mathematics and allied subjects might have been greatly overstated by my Italian friend. I was too busy that Friday to do anything about it, but came home to tell Jonathan that there was a chance. Monday morning I had a little time and rang this man in Milan, who listened with impatience to my explaining of what mathematics my son had done and what he had not done. And then he interrupted me and said, "Look, there is just one requirement for young people to join my outfit. Can your son read books in English?" For, of course, at that time all computing books were effectively still in English. So Jonathan left home on Tuesday and started work on Wednesday, in a most happy relationship with the firm. He went back to it during his time in university and afterwards also for extended periods, quite apart from the initial year he spent there. He is now married and they have two boys.

Our third child, Elizabeth, went from her "A" levels to train as a cook, then worked as a chef for quite a while. Next she went into insurance, passed all the exams very speedily while working, and then decided to go to university, after a break of five years. But she became the one academic in the family, and is now a lecturer at Edinburgh University.

The fourth one, David, did a year's work in ICI's Research Laboratories at Runcorn before going to university to read chemistry. He got a Ph.D. but, much earlier in life than I, left the academic life for a managerial career. The youngest, Deborah, went to the same firm in Milan where Jonathan had been, also for a year, before reading mathematics and then computing at university, and then worked as a software engineer. She is now married, and they have a boy and a girl. Incidentally, like her brother, she went to Italy aged 18 without any Italian, but this bothered neither of them for any length of time before they learnt it. Incidentally, I myself have never been any good at languages other than my first language, German, and English. But this has not affected at all the addiction I have for travel. This started in my boyhood from which I remember holidays, the topography of mountains, walks, and railway journeys far better than anything else. The limitations on travel that the war imposed (except for my enforced trip to Canada, and a little travel on the job, especially to Snowdon) only strengthened the appetite. With my parents in the United States, of course I wished to go there as often as I could manage it, at the then high fares. My first trip in 1946, and the second one, with Christine, in 1947 received no financial support, but later, during sabbaticals, in 1951 (three of us), 1953/4 (four of us), and 1960 (all seven) there were jobs in the U.S.A. which indirectly paid for the travel costs. The 1947 skiing holiday and the 1948 summer holiday (when both Christine and I received travel grants to attend the Zürich Assembly of the International Astronomical Union) ended pleasure travel to the Continent for fifteen years, while the family was growing.

However, business travel to meetings increased rapidly in number, and I rarely said "No" to invitations. Indeed it was said about me that I would come to any meeting, provided it was more than 3000 miles away. Throughout the fifties, this gave me numerous opportunities to visit my parents. I enjoyed every distant country I visited. As regards the U.S.A., I like being there as long as it is not New York. My general dislike for cities is even stronger there than here and fully shared by Christine. But we have enormously enjoyed the hills and the small townships of New England and upstate New York, the historic and scenic aspects of

Virginia, and the National Parks, especially in the Western part of the country. It is a matter of regret to me that, thanks to television and films, so many people in this country believe the United States consists of freeways and apartment blocks, when it is in fact so much a country of one family houses and small towns and locations. Few people here appreciate that one can live on a holiday as we did in 1983, in a small place in northern California consisting of a gas station, a tavern, a store, and a few small houses in the woods (one of which we had rented), all served by a single very minor road.

In Australia we have many good friends. Christine has been there only once, I many times, and each occasion has been greatly enjoyed. We have had wonderful hospitality, friendliness, scenery, and generally a splendid time on our one joint trip to New Zealand.

India is another country of which I am particulary fond. I have already described my very brief trip in 1968 when, on the invitations of that eminent scientist and citizen Vikram Sarabhai I visited the nascent Indian space effort in his city of Ahmedabad, and was greatly impressed. (In 1985 I was invited to give a lecture in his memory there and was delighted to be able to do so.) Next, in January 1973 I led a mission of the Ministry of Defence to strengthen research and development co-operation with India. I visited many of their scientific and technical centres with excellent people and facilities, making it a most interesting and enjoyable visit. I fell in love with Bangalore then, and was especially intrigued by the Army's clothing research establishment near Agra which supports the Indian Army in their confrontation with the Chinese in Ladakh, at extreme altitudes. But whereas the Chinese lines of communication are over the gently rising dry Tibetan uplands, the Indian ones have to cross snow rich mountains and deep valleys and are generally unusable in winter. It is a major task to clothe people at over 4000m altitude suffering from altitude deterioration, so that they are combat ready and not disabled by frost bite.

After a gap of over a dozen years I returned to India with Christine this time, on a British Council arranged and superbly organized visit to numerous scientific and technical institutions all over India. It was a splendid and useful five weeks' visit in the Christmas vacation 1985/6. We made new friends and met old ones especially in Delhi and Bangalore, where Dr. Radhakrishnan gave us a home from home in the Raman Research Institute. It was most impressive to observe the splendid progress India had made in her science, in her space effort, but above all in establishing a dependable supply and distribution of food, while becoming aware of ecological imperatives. Only two years later (1987/8) we returned to India at the invitation of the Indian Humanists. This time we were accompanied by our eldest daughter. The active work of Indian Humanists amongst women in Bombay was fascinating, as was the outstanding, effective and far reaching work of the Atheist Centre in Vijayawada. Of course we also took the opportunity to do some excellent sight seeing and to visit our scientific friends in Bombay and especially in Bangalore. Another visit to India, albeit a very brief one, is planned at the time of writing.

Our younger son David and his wife now live in southern Chile, near Puerto Montt, where he is active in the growing farming of salmon. This gave Christine and me another excuse to travel, and we had three weeks in March and April 1989 in that remote but most beautiful area. Sombre lakes with wooded shores are overtopped by huge snowcapped volcanoes.

Nearer home, in Continental Europe, it is the Alps, and especially the Western Alps and the Dolomites that have attracted us on innumerable occasions, as have the beauty, the art and the antiquities (perhaps especially the Etruscan ones) of Italy, and of southern France (we dislike its northeast though we greatly enjoy Alsace). But of course we have travelled elsewhere in Europe and outside, and have virtually always greatly enjoyed it.

But we are also very fond of what can be found in this country. For walking we love the Yorkshire Dales and Moors, and the North Pennines, what we know of Scotland, and very much like Devon. But best of all are the numerous antiquities spread all over the country, and so well cared for. One can break any journey almost anywhere, and find something fascinating to see, be it a ruined abbey (an evocative sight of which this country has almost a monopoly) or Roman remains or a medieval house, or whatever.

One of my pleasures is to read old travel guides. They are often much better, especially in their maps, than new ones, and have much less of a tendency to drive the tourist to seek out only the most popular sights. I take particular pleasure in locating a little known Etruscan antiquity in the Italian countryside with the aid of an 1893 Baedeker.

Maybe all my career was a mistake. I should have become a travel agent.

Index

Printed and bound by CPI Group (UK) Ltd, Croydon, CR0 4YY

03/10/2024

01040418-0001